Use of Workers' Compensation Data for Occupational Injury & Illness Prevention

David F. Utterback and Teresa M. Schnorr, Editors

DEPARTMENT OF HEALTH AND HUMAN SERVICES

Centers for Disease Control and Prevention

National Institute for Occupational Safety and Health

DEPARTMENT OF LABOR

Bureau of Labor Statistics

May 2010

Revised August 2010

This document is in the public domain and may be freely copied or reprinted.

Disclaimer

Sponsorship of the Workers' Compensation Data Use Workshop and these proceedings by the National Institute for Occupational Safety and Health (NIOSH) and the Bureau of Labor Statistics (BLS) does not constitute endorsement of the views expressed or recommendations for the use of any commercial product, commodity, or service mentioned. The opinions and conclusions expressed in the presentations and report are those of the authors and not necessarily those of NIOSH or BLS. All conference presenters were given the opportunity to review and correct statements attributed to them within this report.

Recommendations are not final statements of NIOSH or BLS policy or of any agency or individual involved. They are intended to be used in advancing the knowledge needed for improving worker safety and health.

Ordering Information

To receive documents or other information about occupational safety and health topics, contact NIOSH at:
Telephone: 1-800-CDC-INFO (1-800-232-4636)
TTY: 1-888-232-6348
Email: cdcinfo@cdc.gov
Or visit the NIOSH Web site at www.cdc.gov/niosh

For a monthly update on news at NIOSH, subscribe to NIOSH eNews by visiting www.cdc.gov/niosh/eNews.

DHHS (NIOSH) Publication No. 2010 – 152

May 2010

SAFER • HEALTHIER • PEOPLE™

Foreword

Tracking health outcomes and their related behavioral and environmental factors is a vital public health function. The National Academies has urged greater use of occupational injury and illness tracking data at the national level to identify priorities, focus resources, and evaluate prevention program effectiveness.

In September 2009, the National Institute for Occupational Safety and Health (NIOSH) and the Bureau of Labor Statistics (BLS) partnered with the National Council on Compensation Insurance (NCCI) and the Washington State Department of Labor and Industries, Safety and Health Assessment and Research for Prevention (SHARP) program to sponsor a workshop on the use of workers' compensation data for occupational safety and health surveillance. Workshop participants came from academia, insurance companies and associations, self-insured corporations, labor unions, and state and federal government.

Prominent researchers and stakeholders described and discussed potential use of workers' compensation data to track occupational injuries and illnesses, assess their burden, and identify innovative ideas for intervention. Panels concentrated on methods, the roles and perspectives of different stakeholders, and the factors that drive changes in incidence and cost. Opportunities and next steps were discussed in general sessions.

These proceedings serve to inform the many stakeholders who did not attend the workshop. More importantly, these contents form a basis for continuing a dialogue on the use of workers' compensation data to track occupational injuries and to identify opportunities for protection of workers' health and well-being.

John Howard, M.D.
Director
National Institute for Occupational Safety and Health
Centers for Disease Control and Prevention

Keith Hall, Ph.D.
Commissioner
Bureau of Labor Statistics

Table of Contents

Foreword iii
Introduction vii
Background vii
Acknowledgements ix
Welcoming Remarks 1
Using Workers' Compensation Data for Occupational Injury and Illness Prevention 3
Safety & Health Assessment and Research for Prevention (SHARP) Program 5
Reducing Occupational Injury - The Value and the Challenge of Determining the Burden 11
A Brief History of Economists' Research on the Effect of Workers'
Compensation on Safety and Health 15
The Contribution of Workers' Compensation Research to Public Health 19
Reconciling Workplace Injury and Illness Data Sources 23
Methodological Challenges in the Liberty Mutual Workplace Safety Index: 29
Harmonizing Existing Databases Counting Workplace Injuries and Illnesses 37
Overview of an Insurance Carrier's Service Sector Data 43
Workers' Compensation Data Utilization in Injury Prevention Research at the
Liberty Mutual Research Institute for Safety 49
Using Employer Records – Pitfalls and Opportunities 55
Multi-Agency Data Matching to Detect Suspected Uninsured Employers:
Research Impacts Policy 63
State-based Occupational Injury and Disease Surveillance 73
Managing Prevention with Leading and Lagging Indicators in the
Workers' Compensation System 83
Benchmarking and Performance Measurement for Governments 89
Self-Insured Experience with Workers' Compensation 93
Using Workers' Compensation Data: The Move from Lagging to Leading Indicators 95
Past, Present, and Future Uses of Some Workers' Compensation Data 97
Differences among State Workers' Compensation Laws and Regulations 105
National Averages of Employee Benefits and Employer Costs for
Workers' Compensation 109
Learning from Workers' Compensation Claims Triangles 115
Identifying and Tracking Trends in Workplace Injuries and Illnesses –
Opportunities and Challenges in Using Workers' Compensation Rating Bureau Data 119

Identifying Vulnerable Populations in Workers' Compensation Data:
Limited English Proficiency Workers and Temporary Agency Workers 125

How to Make Interventions Work: An Insurance Perspective 131

Narrative to Accompany "Barriers to Reporting" 135

Comparing Lost Work Days under Workers' Compensation and Short-term Disability,
Evidence from IBI's Disability Benchmarking Data 141

Linking Workers' Compensation and Employment Security Data for Occupational Health
and Safety Surveillance 145

Reconfiguring a Workers' Compensation Database for Epidemiologic Analysis 149

The Use of Workers' Compensation Data to Identify and Track Workplace Risk and the
Effectiveness of Preventative Measures 155

Data Linkage for Prevention: Traumatic Injuries in Construction 159

Workers' Compensation Coverage by State 165

Summary of Workshop Discussion: Occupational Health and Safety Surveillance
Using Workers' Compensation Data 169

Workshop Participants 175

Acronyms and Abbreviations 181

Using Workers' Compensation Data for Occupational Injury & Illness Prevention 185

Introduction
David F. Utterback, NIOSH

A National Academy of Social Insurance annual report, *Workers' Compensation Benefits, Coverage, and Costs, 2007*, states that workers' compensation insurance covered more than 131 million U.S. workers at a total cost of $85 billion to employers in 2007.[1] Total private insurance coverage accounted for nearly 60% of this total while state managed funds provided about 17%, Federal funds provided about 5%, and self-insurance accounted for more than 18%, respectively.

Total economic and social burden of occupational injuries and illnesses can only be roughly estimated.[2] Uncertainties are due to many factors such as workers receive only a portion of normal wages through compensation, occupational illnesses are frequently not compensated, and insurance data are fragmented and protected for proprietary and personal identification purposes. No central repository for WC claims information exists.

The Workers' Compensation Data Use Workshop was convened to discuss opportunities for collaboration in the analysis of WC data in order to help reduce the risks of occupational injuries and illnesses. Stakeholders from private insurance carriers, insurance associations, self-insured corporations, academic institutions and government agencies participated. Presentations described differences among state laws, proper interpretation of common industry terms, proprietary interests in insurance data, public release of internal analyses, and methods for linking WC data with other health and employment data.

Background

In every State except Texas, nearly all employers are required to have WC insurance for: (1) payment of medical expenses resulting from occupational injuries and some specified occupational illnesses; and (2) partial replacement of workers' lost wages. Each State legislature and the District of Columbia establish workers' compensation requirements with significant variations. For example, states vary in the coverage of compensable occupational illnesses, levels of payments for partial and total disability, both temporary and permanent, and the minimum days away from work to qualify for wage compensation. In many States, employers with small numbers of employees and other groups, such as farm employers, are exempt from coverage requirements.

All WC insurance programs use their data primarily to pay claims to medical providers and disabled workers. Insurance carriers use an array of proprietary data systems. Most WC insurance carriers are private entities. Partial state funds exist in 21 States and exclusive state insurance programs are found in North Dakota, Ohio, Washington, and Wyoming. There are typically many private carriers in each State. Large employers in the U.S. are often self-insured under regulations established by each State. Carriers are required to provide government agencies with claims information that is used for administrative purposes such as oversight, hearings for adjudication of disputes and other matters.

For public health purposes, WC data for acute injuries are far more complete and representative of population risks than are occupational illness data. Some investigators have used limited WC data to estimate the frequency, magnitude, severity, and cost of compensated injuries and to examine trends over time.

Combining WC carrier data would permit better analysis and tracking of occupational injuries and some diseases. (In public health, use of these tracking systems is called surveillance. See box for a definition.) Health scientists, economists, and others could use the larger, combined data sets for more informative analyses of trends in incidence and costs, identification of health hazards associated with new technologies, evaluation of injury and illness prevention program effectiveness, and to provide employers with information needed to protect a most valuable asset – their workforce.

Yet, combining WC injury and illness data is a major technical challenge. Insurance carriers manage very large data sets in integrated, proprietary systems. Some insurance organizations like NCCI routinely collect standardized data from insurance carriers in many States but their contracts restrict data use to issues directly related to estimating rates for establishments within industries while requiring the protection of the proprietary interests of data contributors.

Although some standardized data coding systems are available, such as Occupational Injury and Illness Classification System (OIICS) and North American Industrial Classification System (NAICS), they are not used by all. Additionally, various State rules on compensability, such as minimum number of days away from work to qualify for wage replacement, present obstacles to harmonizing and interpreting data. Data on WC claims from a single State might present fewer challenges yet the analytical results may not be nationally representative.

Analyses of WC data are described in some articles in this publication. Several have resulted in new knowledge for cost management and hazard controls. For example, the Bernacki document herein describes an intervention program for musculoskeletal disorders and describes WC cost control data for a self-insured entity. The Washington SHARP reports refer to the use of WC data to identify needed interventions.[3,4]

We hope that the information in this document will promote greater collaboration for the analyses of WC that will benefit workers, employers, and the U.S. population as a whole. The summary of the workshop begins on p. 165.

Definition of Public Health Surveillance

The U.S. Centers for Disease Control and Prevention (CDC) defines public health surveillance as the "ongoing systematic collection, analysis, and interpretation of health data essential to the planning, implementation, and evaluation of public health practices, closely integrated with the timely dissemination of these data to those who need to know.

References

1. Sengupta I, Reno V, Burton JF (2009) Workers' Compensation: Benefits, Coverage, and Costs, 2007, National Academy of Social Insurance, Washington DC.

2. Leigh PJ, Waehrer G, Miller TR, Keenan C (2004) Scand J Work Environ Health 30(3):199–205.

3. Washington State Department of Labor and Industries, A Guide to Preventing Dermatitis while Working with Advanced Composite Materials, http://www.lni.wa.gov/Safety/Research/Dermatitis/files/acm.pdf, Accessed on March 5, 2010.

4. Washington State Department of Labor and Industries, Scald Burns in Restaurant Workers, http://www.lni.wa.gov/Safety/Research/Files/RestaurantScaldBurns.pdf, Accessed on March 5, 2010.

Acknowledgements

We greatly appreciate the many contributors to the workshop planning and production. Proceedings articles were provided by each workshop presenter. General discussion group leaders were Barbara Silverstein, SHARP, Emily Spieler, Northeastern University, and Letitia Davis, Massachusetts Department of Public Health. Teresa Schnorr, NIOSH joined Barbara Silverstein, SHARP, and Tom Leamon, of Harvard School of Public Health to serve as Co-Chairs of the workshop. The organizing committee also included Ben Amick, Steve Hudock, John Ruser, Rene Pana-Cryan, Peg Seminario, John Sestito, David Utterback and Steve Wurzelbacher. Les Boden, Seth Seabury, Tom Leamon, Peg Seminario, Nancy Stout, Rene Pana-Cryan, Steve Wurzelbacher, Ben Amick, and Janie Gittleman were Panel Leaders. Notetakers at the meeting were Tim Bushnell, Dawn Castillo, Steve Hudock, Rene Pana-Cryan, John Sestito, Kerry Souza, and Steve Wurzelbacher. Organizational assistance was provided by John Gallahan, Tanya Headley, Joseph Leuliette, and Kerry Souza. Many staff at BLS helped host the workshop. Desktop publishing of the proceedings was completed by Greg Hartle and Aaron Geraci. Cover design by Jose Lainez Kafati (photos by Kate Sumbler / ktpupp at Flickr.com, David Clow / David Clow - Maryland at Flickr.com, St Stev / St Stev at Flickr.com, George Wu / GeoFX at Flickr.com). We are grateful for all contributions.

Welcoming Remarks

Commissioner Keith Hall, Bureau of Labor Statistics

I would like to welcome all of you to the Bureau of Labor Statistics and to this workshop on using workers' compensation data for injury and illness prevention. The topic of this workshop addresses a mission vital to the Department of Labor and the other organizations represented here. That mission is to ensure that every worker returns home from work as healthy as when they left home.

The Bureau of Labor Statistics (BLS) supports this mission by supplying workplace injury and illness data that are widely used to identify high risk workplaces and to help target workplace safety and health interventions. The BLS provides national and State level safety and health surveillance information from the Survey of Occupational Injuries and Illnesses and the Census of Fatal Occupational Injuries.

While these BLS data programs form the underpinning of our national workplace safety and health surveillance system, other data sources, including workers' compensation, can provide vital complementary information. These data can supplement the BLS data with richer epidemiological information on the factors causing or associated with injuries and illnesses. They can provide better information about long run outcomes. And, these data may identify cases that are not captured by the BLS survey, perhaps because they are outside the Survey's scope.

However, there are challenges in using data such as workers' compensation for injury and illness surveillance. These data are not available nationally and they vary from jurisdiction to jurisdiction in the injuries and illnesses they cover. They may be limited in scope, for example, to only those injuries that last a minimum number of days away from work. And, law or regulation may restrict who can access them. So, these data can complement, but not replace the BLS national workplace safety and health surveillance system.

This workshop is about exploring the ways that workers' compensation data can add value to injury and illness prevention and ways that the limitations of these data can be overcome. I commend the organizers and participants on the workshop's excellent and wide-ranging agenda of topics. With the level of expertise of those assemble here, I expect much useful information to come out of this workshop.

I hope all of you leave here with a deeper understanding of how workers' compensation data can achieve our joint mission to protect workers. I wish you a productive meeting.

Using Workers' Compensation Data for Occupational Injury and Illness Prevention

Director John Howard, National Institute for Occupational Safety and Health

Good morning ladies and gentlemen, and welcome to what I hope will become an annual conference bringing together the occupational injury and illness prevention community and the occupational injury and illness compensation community for the purpose of identifying ways that we can work together to achieve what we are all working toward—a safer, healthier, and more secure American workforce.

This is a workshop rich in experts from every field of prevention and compensation, and the National Institute for Occupational Safety and Health is grateful for your attendance. NIOSH is also grateful for your commitment to bringing all of us to a place of fuller understanding about how workers' compensation data can inform workers' prevention decision-making; and how workers' prevention activities can inform workers' compensation loss control efforts.

In the late 1990s in California—and I'd like to pause to acknowledge the presence today of John Duncan, the Director of the California Department of Industrial Relations—I saw that many of the causes of worker injury and illness were not covered by existing standards (and likely would never be) and asked California insurers to partner with the Cal/OSHA Consultation Service in better serving our mutual clients for the benefit of California workers. I hope that that sense of mutuality can be achieved again insurer by insurer and state-by-state.

The injury prevention and injury compensation communities are both facing traditional challenges and are facing some daunting new challenges like:

• An aging workforce;

• An increasingly obese workforce with all the attendant medical manifestations of excess weight; and

• An influx of war veterans entering the workforce after suffering from internal injuries in Iraq or Afghanistan that, but for modern military medicine, would have resulted in mortality previously, and will now complicate any workers' compensation claim should they become injured on the job.

The National Institute for Occupational Safety and Health stands ready to facilitate deliberations and workshops like this one, data exchange interactions and collaborative programs that will lead to a broader understanding of our mutual interests in partnering and our respective concerns about doing just that.

I wish each of you safe, healthful and secure work, but most of all, I wish you a very successful workshop!

Thank you.

Safety & Health Assessment and Research for Prevention (SHARP) Program

Barbara Silverstein, Washington State Department of Labor and Industries

Welcome & Introduction

This conference was called to extend the dialogue about existing data that can be used for occupational safety and health surveillance. Specifically, can workers' compensation (WC) data be used to augment what we know from the Survey of Occupational Injury and Illness (SOII) conducted by BLS? WC and surveillance are not obvious partners. There are divergent interests. Coverage differs among states. Most workers' compensation carriers are not in the business of sharing data for research purposes. Most states have multiple carriers. There is no national repository for detailed workers' compensation data for all states. NCCI and WCRI have some data for some but not all States. It is easier to track acute traumatic events in WC systems than occupational illnesses. Nonetheless, workers' compensation data can be used to estimate the (1) magnitude, (2) severity, (3) cost, and (4) frequency of many injuries, and to look at trends over time. Illnesses are not always as easy to identify in workers compensation data, but we are getting better at that. In Washington State, stakeholders pay attention to trends in workers compensation – it means lives, but it also means money, and it can be used for prevention priority setting purposes; and it can be used to augment (not replace) what we know from BLS.

A brief discussion of how we use WC data for occupational health surveillance and priority setting in Washington State may encourage similar uses elsewhere. Washington State is unique in a number of ways. Washington is the only state where the labor department has both a state OSHA plan and an exclusive workers' compensation system. This permits some unique opportunities to use data from both programs. Business and labor both are represented on Department of Labor and Industries (DLI) advisory committees. SHARP is the only occupational safety and health research group that is located within a state labor department.

One of the things we've done in SHARP is to examine how we can best prioritize the kinds of research we do given "what" our mission is and "where" we are located. Ninety percent of all workers' compensation claims, all time-loss days and costs are within seven different categories (Table 1).

SHARP uses a prevention index (PI) to help prioritize information for action. The PI is constructed by rank ordering all industries by claims incidence rate and by incident count and then averaging the two ranks (PI = (Incidence rank + Count rank)/2). Different prevention strategies may be used depending on where an industry is ranked (Table 2).

Trucking: a high risk industry

Using incidence rates and incident counts in the prevention index, we've focused on selected industry sectors, rather than focusing on selected injury or illness conditions-directed research.

When we used the prevention index to rank order industries for potential intervention, we had already begun work with construction, logging was small and both of those industries were the focus of attention for the state OSHA program (Table 3). Virtually no work (either enforcement or consultation) was being performed in the trucking industry. The unspoken assumption was that there was other government agencies that regulated trucking, primarily related to road safety, so they would cover worker safety. However, that assumption was wrong.

A good example of this is our work in the trucking industry (www.keeptruckingsafe.org). We have a very active TIRES (trucking injury reduction emphasis on safety) labor-management steering committee that helps guide our surveillance and research activities. While this industry is heavily regulated in terms of highway safety, there has been little focus on occupational safety and health for drivers. We conducted needs assessment surveys among trucking employers and truck drivers. We also do workers' compensation case follow-up and root cause field investigations; develop educational materials and pilot prevention activities in select companies.

Safe Patient Handling

Another example of using workers' compensation data was a request by a state legislative committee to conduct a review of reasons for high rates of WC claims in health care. Claims incidence rates for nursing homes have been high for a long time but were declining. Incidence rates for hospitals were lower but had begun to rise (Table 4).

A stakeholder committee was formed, with visits to different types of facilities where manual handling injuries were identified as the major area for improvement. During the next legislative session, Safe Patient Handling (SPH) legislation was enacted for hospitals with financial incentives for purchasing patient lifting and moving equipment as well as requiring a joint SPH committee, conducting evaluations and full implementation within four years. A statewide SPH steering committee was established by stakeholders (labor, management, SHARP) to assist in implementation with an active website (www.washingtonsafepatienthandling.org). SHARP evaluation of this legislative intervention includes the following:

- Compare WC rates over time for hospitals (legislation) vs. nursing homes (no legislation)
- Survey hospital management and staff regarding implementation
- Compare concordance between staff and hospital views on implementation (H_1: Those with most concordance will have lower injury rate)
- Compare Washington State hospitals to those in another state without legislation (Idaho)
- Compare administrative data: Departments of health, labor, revenue, and employment security

SHARP continues to monitor program elements, including using business and occupations tax credits for purchase of equipment, site inspections by the Health Department and workers' compensation claims rates by the Department of Labor and Industries. Additionally, SHARP is comparing implementation in 4 Washington hospitals (legislation) with 4 Idaho hospitals (no legislation) using both quantitative and qualitative approaches.

Table 1. Washington State Fund Compensable Claims Costs and Time loss Days for 7 Injury Types, 1998-2004.

Type of Claim	% of all Claims	% of all Costs	% of all Time Loss days
Neck, Back, Upper Extremity MSDs	42.3%	45.2%	49.3%
Struck By/Against	15.6%	12.9%	12.7%
Fall on Same Level	9.1%	10.1%	10.7%
Lower Extremity MSDs (LE)	7.6%	6.5%	6.2%
Fall from Elevation	6.6%	10.3%	10.5%
Motor Vehicular	2.9%	4.7%	3.8%
Caught in/under/between	2.4%	2.2%	1.8%
Other	13.4%	8.1%	4.9%

Table 2. Prevention Index Strategies

Rate and Count Combination	Intervention Strategy
High Rate with High Count	Industry-wide approach with enforcement, consultation and education/outreach
High Rate with Low Count	Risk concentrated in small industry. Focused inspection approach may be appropriate
Low Rate with High Count	Risk in large industry; lots of people. Likely no single workplace at high risk: education campaign
Low Rate with Low Count	Minimal resources needed unless complaints or unique injuries/ hazards emerge over time

Table 3. Prioritizing Industries by the Prevention Index

Industry	UEWMSD	Fall Same Level	Fall-Elevation	Caught	LEWMSD	Motor Vehicle	Struck	Average PI
Logging	18	1	5	2	5	4	2	5.3
Building Construction	1	2	1	14	2	22	1	6.1
General Freight Trucking	8	3.5	6	13	4	1	9	6.4
Residential Construction	5	7	3	33	3	38.5	3	13.2
Specialty Freight Trucking	9	9	7	45	9	2	14.5	13.6

Note: N=278 industries in Washington State Fund
PI= [Incidence rate rank + Count rank] /2

Table 4. Washington State Fund & Self Insured Compensable Claims Rates for Health Care Workers: OtherPerson, Acute Care Hospitals NAICS 622110

					Compensable Claims Rate per 10,000 FTE				
Injury Type	Year	2000	2001	2002	2003	2004	2005	2006	2007
All Compensable Claims-Other person ALL		115	109	112	97	108	95	120	121
All Comp-other person- Hospital		158	146	135	132	113	109	110	105
All Comp-other person-Nursing Homes		256	313	268	270	209	186	191	158

Figure 1. SHARP Upper Extremity Study

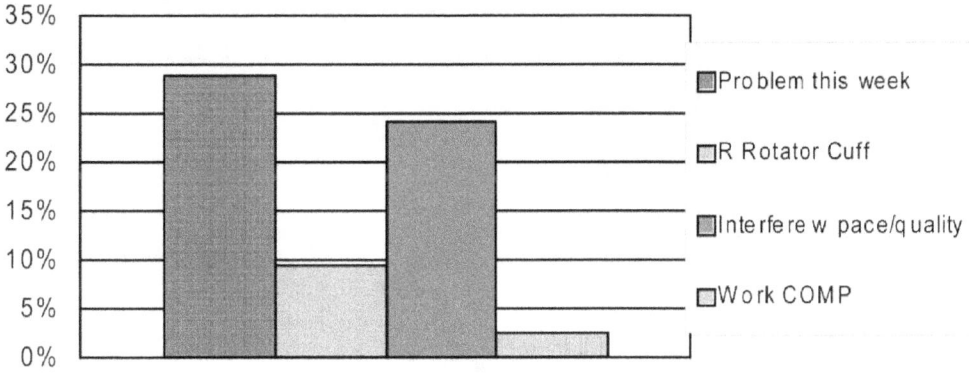

Work-related Asthma

Another example includes using WC data to identify industries with a potential increased risk of asthma and dermatitis. While SHARP also maintains a provider reporting program for work-related asthma, we have been able to identify clustering in certain industries using the WC data. Specifically, we identified high incidence in the collision repair industry. This is an industry of small employers that had received very little state OSHA attention. SHARP researchers, in collaboration with the industry association and researchers at UNC-Chapel Hill, were able to determine high diisocyanate absorption from respiratory and dermal exposures. This has lead to further research on different gloves. WC claims will continue to be monitored as different control measures are implemented.

WC underestimates injuries and illnesses

While workers' compensation data can be used for surveillance purposes, this likely produces under-estimates of prevalence and incidence. In Washington State this is shown through separate studies in establishments wherein injury and illness data are compared to workers' compensation claims. For example, in a SHARP study of upper extremity disorders in manufacturing and health care facilities (n=660), prevalence of pain and clinical cases of rotator cuff tendinitis was much greater than WC case prevalence for the same workplaces (Figure 1).

Workers compensation data can also be combined with unemployment data to better understand the burden of occupational injuries and illnesses for workers. For example when comparing earnings after filing a claim for carpal tunnel syndrome or upper extremity fracture, it is clear that workers with CTS generally do not recover their pre-claim wages even seven year after claim filing whereas there is a rapid return to full wages for those with fractures (Figure 2). These kinds of WC trend analyses can be used to focus prevention efforts.

Concluding remarks: Surveillance is really an issue about what is "under the surface." Surveillance provides us with a picture of the broad spectrum of what is happening. It prompts us to perform more basic research to provide explanations for why "it" is happening, and "what" to do about it."

Figure 2. Median quarterly earnings as percent of injury quarter (Foley, M, Silverstein B, Polissar, NL, AJIM 2007)

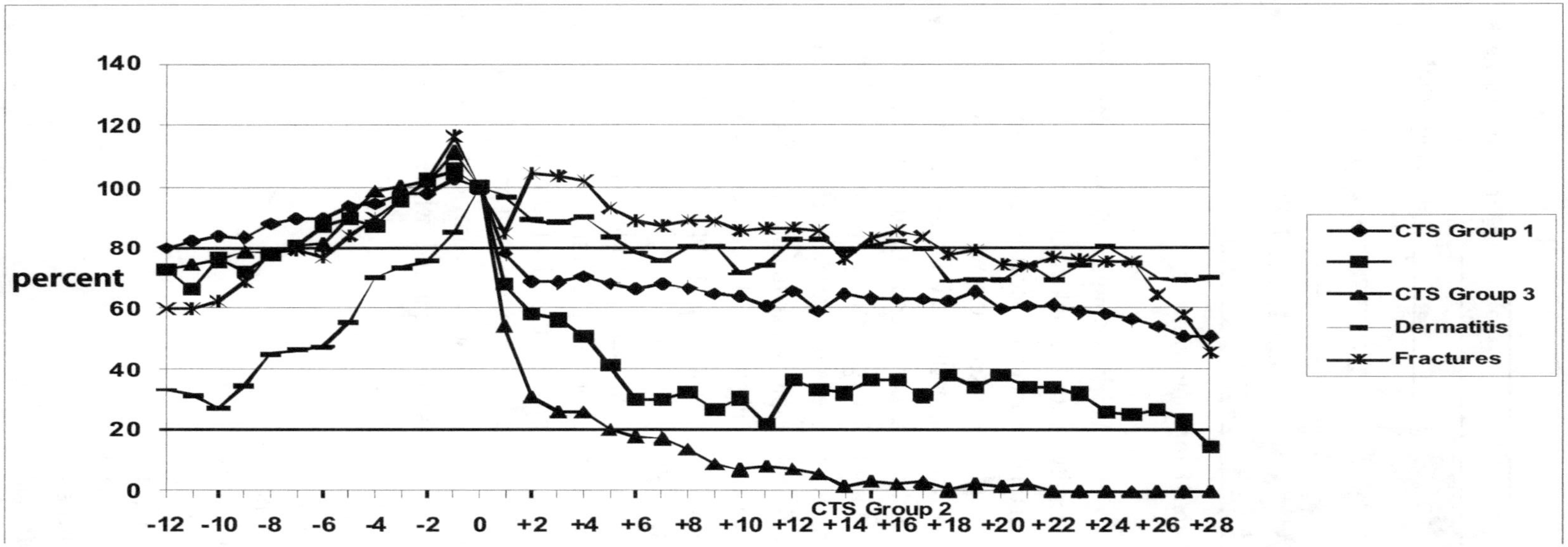

Reducing Occupational Injury - The Value and the Challenge of Determining the Burden

Tom B. Leamon, PhD, Adjunct Professor, Harvard School of Public Health

Occupational Injury and Interventions

There appears to be a palpable lack of public concern for the immense burden which occupational injuries impose on American society and individual enterprises. This, despite the fact that the direct and indirect cost burdens arise from the pain and suffering of individual workers, may result from a lack of information on the burden per se. This lack of good data is by no means solely an American problem. The current data produced by the International Labour Organization (ILO), (and well known by that institution), indicate how serious an issue this is; with Pakistan reporting fewer fatal injuries than Singapore and India reporting fewer fatalities than Hong Kong. Without appropriate data it should be expected that appropriate research and intervention resources will not be available to reduce this burden.

In the United States, where the number of occupational injuries reported exceeds the number of new cases of disease reported by a factor of more than the 13 to one, a lack of data on injury obscures the need for research and intervention into injury prevention. While this observation on injury vs. disease does not accommodate the very significant number of long latency disease cases which arise from workplace exposures, it is clear that the resources, both intellectual and financial, devoted to reducing injury are not allocated to reflect the relative significance of each. This is not to say that the resources devoted to avoiding occupational disease should be reduced -- for the evidence is that the current resources have made, and continue to make, significant improvements to workers health. Instead, it is a cry for the allocation of more resources appropriate to the burden of injury placed upon individual workers, their employers and the broader society. Such a plea is unlikely to be heard without surveillance data on the burden of injury.

The unacceptable lack of resources devoted to reducing occupational injuries can be readily seen by charting the number of Schools of Public Health in the United States with comprehensive occupational safety programs. An analysis of School web sites, identified by the Association of Schools of Public Health, shows that of the many schools active in occupational safety and health there is not a single one claiming such an occupational safety program. While a significant proportion of the Schools, (but not even a majority) identify safety in their course or activity lists, none appear to address occupational safety in an appropriate manner – instead, topics included under this rubric include violence (spousal, hand gun etc), youth, bicycles, rural & agricultural exposures and automobiles - including collision biomechanics. Analysis of individual web sites reveals that activities as varied as mental-health economics, various HIV interests, drug abuse, obesity and tobacco products were included under the "safety" banner.

The Current Approach to Reducing Occupational Injury

In the competitive environment currently found in enterprises, safety and health interventions are likely to compete for scarce resources with other priorities and consequently the absolute size of the burden is of significance. Hitherto, the most significant attempt to determine the burden is perhaps the WHO/Harvard initiative - the Global Burden of Disease, which attempts to measure the burden by the use of Disability Adjusted Life Years (DALYs). There is a significant and critical literature, concerning methodological challenges to this metric.

Of particular concern to the present commentator, besides the method of determining disability weights is the practice of developing monetarized derivatives in order to rank the seriousness of the various sources of disability. The popular, and well-intentioned, approach to establishing the seriousness of this issue by hypothesizing an ever expanding view of the social consequences of occupational injuries and illnesses, (which generates a colossal, but hypothetical, monetarized value), may be counter productive. Such an approach appears to depend on attracting the interest of a super-enterprise party and then waiting for a "deus ex machina" intervention to make the improvement. In other circumstances, for example several road safety initiatives, significant burdens measured by DALYs have attracted governmental interest and have produced legislation which, when coupled with enforcement, have led to safer circumstances. However in the case of occupational injury a concern is the situation where the estimated "societal cost" is very substantially larger than the actual incurred costs. In this circumstance, given the well accepted huge variability in these estimated costs, a slight error in this estimated component may totally eclipse the actual costs borne by the appropriate party. If this were not a sufficient challenge to those responsible for the introduction of interventions, an even more serious problem is that any savings by reductions in this estimated component are not realizable by those responsible for the introduction of, and the cost for, the appropriate interventions which is normally the "workplace owner" i.e. the employer. Unfortunately, in the current environment where the expenditures involved in any enterprise intervention must be competitive with other financial demands this approach is likely to fail and the much smaller "green" dollar savings are inevitably likely to receive more attention than the "white" dollar version.

Data Issues

In determining the appropriate measure of the burden there are significant technical challenges to be taken up and many will be discussed in this meeting. Four challenges typically not pursued, but which may be partially addressed by the use of workers' compensation (WC) data include:

"Proportional" Reporting

In terms of reporting, there is simply a wide variation in the understanding of what should be reported. Workers with sharp instruments or glass workers may ignore many minor cuts, and miners with intermittent low back pain may assume this is part of their occupational demand. The wide variability of work environments, from office reception areas, to forestry or fishing in winter also leads to different perspectives on the seriousness, and hence the reportability, of various injuries. In contrast with the Occupational Safety and Health Administration (OSHA) records, WC costs partially overcome this depending as they do on a decision to expend money made by an external and critical payee. Clearly, this is not to claim that this is a more accurate measure of seriousness, but only that it is likely to be less subject to variability than missed time. A similar argument might be extended to the benefits of using data from only the more serious claims, for example those exceeding three or five days.

Parenthetically, the philosophical challenge of developing scales to allow comparisons between risks and, even more appropriately, to allow meaningful legislation to accommodate this immense range of environments is one which should be inherent in future research.

Defining "Occupational"

The practice in some administrative databases to exclude certain categories of exposure to work hazards, including sections of agriculture, self-employment and youth, is well recognized.

Less well researched is the actual definition of an occupation exposure, including the question of work-for-pay or not-for-pay. In a study in a developing economy we found that 42% of injured workers reported injury occurring in the workplace, compared with a 62% response by the same workers to the question of: "Were you hurt while working?".

[The same study found significant differences when injury rates were calculated using the traditional approach of using the number of jobs as the denominator compared with using the number of "full-time equivalents". In a society involving an increase in part-time work, post-retirement work, workers working at home and multiple jobs this is a serious issue which requires research in order to develop appropriate corrections.]

Transient Workplaces
The question of transient workplaces is acute in construction, forestry and other high risk environments. In these environments, workers can be exposed for short periods to high risks, risks which may not be replicated for significant periods. In many of these industries, improvisations to overcome unforeseen difficulties are necessary and are likely to continue to generate acute, but short term risks. New approaches, such as case crossover designs may be needed to determine both the burden and the significance of particular hazards.

Transport vs. Occupational
The widespread practice of breaking out "transport" from "occupational" exposures obscures the seriousness of occupational exposures. This is certainly the case in the ILO figures - especially for those countries in which much, or even most, transport injury is associated with occupational uses. Equally, it should be pointed out, that many so-called manufacturing enterprises in this country are in fact huge transportation businesses, with many workers involved in trucking and the use of regular automobiles in the course of their occupation.

Conclusion
The value of the accurate determination of the Burden of Occupational Injury and Disease lies in the potential facilitation of workplace improvements and the reduction of hazards. The use of WC data, while producing new issues, may be a unique contributor to this process and this potential justifies the continuation of attempts to match the needs of the carriers and their customers, statutory bodies and researchers.

Post-script
Finally, the measurement of the Burden may address the largely overlooked, or ignored, fundamental difference between the non-fraternal twin issues of "Safety" & "Health". In traditional health investigations the role of surveillance is often to identify subtle or concealed risks and relationships. This disease model approach has less value in many traumatic injury exposures which, in themselves, are clearly hazardous. In this case a significant role for a surveillance system is to facilitate interventions, by increasing awareness of the huge burden paid by American enterprises for the pain-and-suffering borne by their workers as a result of workplace hazards.

A Brief History of Economists' Research on the Effect of Workers' Compensation on Safety and Health

John F. Burton, Jr., Rutgers University

Experience Ratings and Safety: The First Sixty Years

Encouragement of workplace safety has been a basic objective of workers' compensation programs since their origin in the United States. John R. Commons, an economist at the University of Wisconsin and the "father of American social insurance," helped design the 1911 Wisconsin workers' compensation program, the oldest state program. A key feature is that the insurance premiums were experience rated. The rationale was provided in Commons and Andrews (1936: 255-56):

> "One company may perhaps take great interest in safety work, while another does not. The former would be a better risk than the latter and is entitled to a lower rate. This allowance is accomplished under a merit rating system. Instead of one flat rate for an entire industry, this system seeks to adjust the rate of each employer to the hazard of his particular establishment.... Neither insurance companies nor state funds have power to compel the safeguarding of machinery, but they can frequently attain the same end by increasing or reducing the insurance rates under the merit system previously discussed."

The contribution of experience rating of workers' compensation premiums to workplace safety was generally accepted for decades after the emergence of state workers' compensation programs. However, The Report of the National Commission on State Workmen's Compensation Laws (1972: 96-97) indicated it was difficult to demonstrate a statistical relationship between experience rating and the safety records of industries and firms, and noted there had been few systematic attempts to evaluate the relationship of workers' compensation insurance rates to safety. The National Commission provided a few tests of the effects of experience rating on safety, of which the most sophisticated was presented in Figure 1. The data compared states on the basis of their generosity of workers' compensation benefits to their relative injury frequency rates. The National Commission concluded: "There does not appear to be a systematic relationship ... between the level of benefits and the safety record in the State." [1]

Experience Ratings and Safety: The Last Thirty-Five Years

Burton (2009) and Burton and Chelius (1997) provide an overview of the studies since 1972 of the relationship between experience rating of workers' compensation premiums and workplace safety. Workers' compensation programs actually rely on two levels of experience rating to promote safety. Industry-level experience rating establishes a pure premium (or manual rate) for each industry (or occupation) that is largely based on prior benefit payments by the industry. The resulting differences in labor costs and prices among industries should shift the composition of national consumption towards safety products. Firm-level experience rating determines the workers' compensation premium for each firm above a minimum size by comparing its prior benefits to those of other firms in the industry. In order to remain competitive, firms have an incentive to improve safety in order to reduce premiums, as postulated by Commons.

---

[1] The National Commission did not report the correlation between the average benefit level and the state injury frequency rate, which is roughly -0.277.

Figure 1. Relationship between State injury frequency rate and State workmen's compensation average benefit, (indemnity and medical), per case, 1968 -1969 policy year

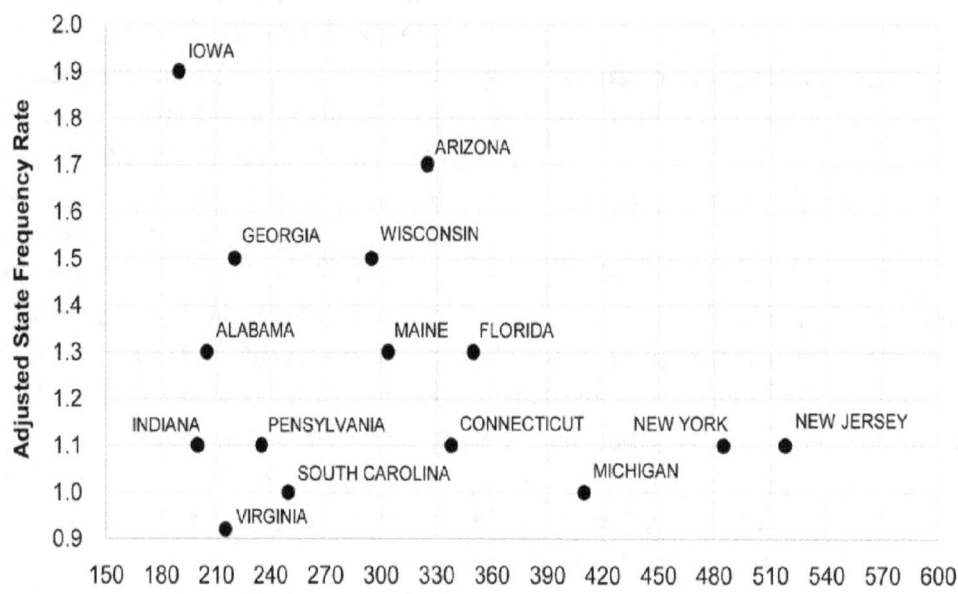

The effects of the workers' compensation program on safety in general, and firm-level experience rating in particular, have been debated by a number of scholars. The essence of the "pure" neoclassical economics approach is that the introduction of workers' compensation will lead to smaller risk premiums in the wages paid to workers and thus reduce the incentives for employers to prevent accidents. Arguably, the increased incentives to safety from experience rating will be entirely offset by the reduction in the risk premium in wages, thus resulting in no improvement in safety from the use of experience rating in workers' compensation.

In contrast, the economists who do not endorse the "pure" neoclassical approach argue that the introduction of workers' compensation with experience ratings should improve safety because in the absence of the program, the limitations of knowledge and mobility and the unequal bargaining power of employees mean that the risk premiums generated in the labor market are inadequate to provide employers with the safety incentives postulated by the neoclassical economists. These economists argue that experience rating should improve safety by providing stronger financial incentives to employers to avoid accidents than the muted incentives provided by risk premiums.

A number of studies provide evidence that should be helpful in evaluating the virtues of experience rating in workers' compensation. However, the evidence is inconclusive. A survey of the literature by Boden (1995: 285) concluded that "research on the safety impacts has not provided a clear answer as to whether workers' compensation improves workplace safety." In contrast, Thomason (2005: 26) asserted that most (11 of 14) of the studies he surveyed found that experience rating improves safety and health and that studies failing to detect the relationship were methodologically weaker than the other studies. Thomason concluded that "Taken as a whole, the evidence is quite compelling: experience rating works." Tompa et al (2007: 91) also surveyed the literature and found that moderate evidence that the introduction of experience rating reduces the frequency of injuries (although the severity may increase) and moderate evidence that the degree of experience rating reduces the frequency or severity of injuries.

Thomason (2005: 27) cautioned that experience rating may, in addition to encouraging employers to improve workplace safety and health, also lead to increased claims management by employers, including the denial of legitimate compensation claims. While the evidence suggests that on net experience rating is associated with improved workplace safety, there are variations among employers in accident prevention efforts relative to claims management efforts.

Further Research Needs
Despite the extensive literature on the effect of experience rating of workers' compensation premiums on safety in recent decades, there are many topics that warrant further study.

First, workers' compensation insurance policies with large deductibles have increased from 2.8 percent of all workers' compensation benefits in 1992 to 14.8 percent in 2007 (Sengupta, Reno, and Burton 2009: Table 6). Employers who have policies with deductibles are, in effect, relying on perfect experience rating up to the amount of the deductible. If so, there should have been a discernable effect of the increased experience rating on workplace injury rates since the early 1990s.

Second, Guo and Burton (2010) found that a substantial portion of the decline in workers' compensation benefits during the 1990s was due to more restrictive eligibility rules enacted by many states during the decade. If so, the reduction in benefits paid by employers should have reduced the incentives to improve workplace safety.

Third, the variations among employers in the effect of experience rating on safety efforts as opposed to resisting legitimate claims has only received limited attention. Of particular interest is whether the increased use of large deductibles has affected the relative importance of these employer responses to experience rating.

References
Boden, Leslie I. 1995. "Creating Economic Incentives: Lessons from Workers' Compensation." In Proceedings of the Forth-Seventh Annual Meeting. Madison, WI: Industrial Relations Research Association.

Burton, John F., Jr. 2009. "Workers' Compensation." In Kenneth G. Dau-Schmidt, Seth D. Harris, and Orly Lobel, eds. Labor and Employment Law and Economics. Cheltenham, UK: Edward Elgar: 235-74.

Burton, John F, Jr. and James R. Chelius. 1997. "Workplace Safety and Health Regulations: Rationale and Results." In Bruce E. Kaufman, ed. Government Regulation of the Employment Relationship. Madison, WI: Industrial Relations Research Association: 253-93.

Commons, John R. and John B. Andrews. 1936. Principles of Labor Legislation. 4th ed. New York: Harper and Brothers. [1st ed. in 1916.]

Guo, Xuguang (Steve) and John F. Burton Jr. Forthcoming 2010. "Workers' Compensation: Recent Developments in Moral Hazard and Benefit Payments." Industrial and Labor Relations Review.

National Commission on State Workmen's Compensation Laws. 1972. The Report of the National Commission on State Workmen's Compensation Laws. Washington, DC: Government Printing Office.

Sengupta, Ishita, Virginia A. Reno, and John F. Burton Jr. 2009. Workers' Compensation: Benefits, Coverage, and Costs, 2007. Washington, DC: National Academy of Social Insurance.

Thomason, Terry. 2005. "Economic Incentives and Workplace Safety." In Karen Roberts, John F. Burton Jr., and Matthew M. Bodah, eds. Workplace Injuries and Diseases: Prevention and Compensation: Essays in Honor of Terry Thomason. Kalamazoo, MI: W.E. Upjohn Institute for Employment Research: 9-36.

Tompa, Emile, S. Trevithick, and C. McLeod. 2007. "Systematic Review of Prevention Incentives and Regulatory Mechanisms for Occupational Health and Safety." Scandinavian Journal of Work, Environment, and Health, Vol. 33, No. 2: 85-95.

The Contribution of Workers' Compensation Research to Public Health

Allard E. Dembe, ScD, The Ohio State University, College of Public Health

Many countries maintain comprehensive national workers' compensation (WC) databases containing information on all occupational injuries and illnesses, benefit payments, and conditions in the workplace associated with the injury or illness. For example, researchers at the Republic of Korea's Occupational Safety and Health Research Institute have databases containing complete WC records for all work-related injuries and illnesses that can be linked with employment data and job records for all employees, along with noncompensation health care and hospitalization records. Similar linked databases exist in British Columbia and other locales internationally. The availability of such comprehensive linked data systems creates the potential to conduct studies using detailed job history, along with WC injury and illness data, to examine a variety of questions, such as the effect of work history and job injuries on the risk for chronic conditions later in life.

Performing such a study in the United States would be quite difficult for a variety of reasons. First, there is no similar comprehensive national database of American WC cases. Another challenge is that most WC data systems are used primarily for administrative processing of WC claims, and thus lack important general health and job-related information. This limits the usefulness of WC data for general public health surveillance purposes. Also, it is generally quite difficult (or impossible) to link WC data with independent health system medical care records or with employer job files. While some federal agencies, such as U.S. Bureau of Labor Statistics (BLS), have initiated efforts to track occupational injury and illness occurrence nationally, the resulting data lacks important detail concerning costs, medical treatment, and the specific employment activities, conditions, and exposures associated with the injury.

There are historical reasons why the United States does not have a centralized federal system for collection of WC data. The initial enactment of WC laws between 1908 and 1915 occurred quickly, in response to employer concerns about tort liability for work-related injuries, Progressive Era labor activism, and highly publicized tragedies, such as the Triangle Shirtwaist Company fire of 1911. The growth of workers' compensation laws in the United States took place without substantial federal government involvement or oversight. As pointed out by Ann Clayton (2003), social welfare programs during that period were considered to be local issues, to be administered by state, county, or local jurisdictions.[1] As a result, there was little effort to coordinate the WC programs or to standardize accident reporting systems.

Labor statisticians were aware of the problems created by this lack of uniform accident reporting requirements among states. In 1908, Frederick Hoffman, the pioneering statistician for the Prudential Life Insurance Company, bemoaned the regrettable "lack of completeness and the absence of uniformity" in state accident reporting systems.[2] As late of 1927, Lewis DeBlois of the National Bureau of Casualty & Surety Underwriters, observed that, "adequate machinery for the collection of industrial accident statistics simply does not exist."[3] In an attempt to fill the gap, several employer-supported organizations began to collect industrial accident information. Voluntary efforts to create a standard accident reporting system for WC were undertaken by the National Council on Compensation Insurance (NCCI) in 1919 and National Safety Council (NSC) in 1924. The BLS, in close collaboration with the NSC, devised an approach in the 1930s for estimating occupational fatalities and nonfatal injuries that drew upon various sources of data including reports by NSC member employers, death certificates, surveys of employers, and information from state WC boards. However, all of these methods had significant gaps and methodological shortcomings. As a result, the United States failed to adopt a complete and comprehensive system for compiling WC data and industrial accident information nationally

Because most WC data was in the hands of employers, private WC insurers, employers organizations (e.g., the NSC) and insurance groups (e.g., the NCCI), WC was rarely used for broader public health purposes. Most WC research historically has been confined to studies that addressed the goals and interests of the system participants; particularly financial studies of cost components within the system. It was never the goal for WC data to be collected or applied to broader public health questions. To the extent that WC data has been applied to injury and illness prevention, the primary aim has been to prevent injuries and illnesses in specific workplaces. The difficulties noted earlier in assembling national datasets from WC records have further limited the ability to address public health questions through the use of WC data.

Nevertheless, there have been occasional instances in which WC data has been used in studies that had significant implications for national public health. Examples include investigations of asbestos-related lung disease in miners and shipyard workers during the 1930s and 1940s,[4] studies of noise-induced hearing loss in steel workers during the 1950s,[5] and studies of repetitive motion disorders among meatpacking workers in the 1980s.[6] In each case, those studies brought attention to broader risks faced by the general public in non-occupational community settings.

Though most WC research is directed inwardly towards WC systems needs and health risks occurring in occupational settings, there is tremendous potential (much of which is still untapped) for using WC data to address larger public health issues. For instance, Englehart et al. (1999) used WC claims data for municipal solid waste workers in Florida not only to identify occupational risks to the affected workers, but also thereby to provide an indication of potential risks for populations proximal to landfills, incinerators and other waste sites.[7] Another recent example of how WC research can be directly relevant to wider public health concerns can be found in the study by Rosenman et al. (2003), who, using WC administrative data along with other sources (physician reports, and indication of WC as primary payer on hospital discharge records), identified cases of work-related asthma associated with the use of common cleaning products.[8]

These examples illustrate the potential benefits that could be achieved by applying WC data to investigate issues having impact on communities and individuals outside the traditionally working environment. The greatest benefit might accrue when WC data is linked to other related data sources such as health care records, employment and job records, and surveillance systems. To be most beneficial, a national data system to comprehensively collect WC data in a way that is publicly accessible by researchers should be established. This will require greater uniformity in collection methodology among states. Although NCCI, the WCRI, and other groups have created databases containing composite WC records from multiple states for research purposes, there is still no national system for compiling WC data from all jurisdictions. These limitations place constraints on the ability of WC research and data to be as useful as possible for public health purposes.

In 2001, Gordon Smith, of Johns Hopkins University, articulated a well-conceived agenda for how WC and related work-injury data can be used more effectively to help achieve national public health objectives. [9] His idea was that occupational health and WC data systems ought to become better integrated with other systems and institutional approaches for promoting public health. Smith identified four areas in which traditional public health potentially intersects with occupational health and WC research: a) surveillance, b) risk factor identification, c) intervention development and identification of control strategies, and d) implementation and evaluation of prevention and control programs. In each of these four areas, WC data and research can play a more prominent and useful role. WC researchers and policy makers need to understand the importance of making their efforts reach beyond the narrow confines of traditional WC and employment settings to be clearly relevant to populations and communities and more directly support national public health initiatives.

References

1. Clayton, Ann. 2003/2004. Workers' compensation: a background for social security professionals. Social Security Bulletin 65(4):7-15.

2. Hoffman, Frederick L. 1908. Industrial accidents. Bulletin of the Bureau of Labor Statistics 78(September):417-465.

3. DeBlois, Lewis A. 1927. Has the industrial accident rate declined since 1913? Presentation to the 1927 annual meeting of the National Bureau of Casualty & Surety Underwriters. Accessed August 30, 2009 at: http://www.casact.org/pubs/proceed/proceed27/27084.pdf.

4. Calhoun, Craig; Hiller, Henryk. 1988. Coping with insidious injuries: the case of Johns-Manville Corporation and asbestos exposure. Social Problems 35(2):162-181.

5. Johnson, Eric Daniel. 1993. Sounds of silence for the walkman generation: rock concerts and noise-induced hearing loss. Indiana Law Journal 68:1011.

6. Dembe, Allard E. 1996. Occupation and Disease. New Haven, CT: Yale University Press.

7. Englehardt, James D.; Fleming, Lora E.; Bean, Judy A.; et al. 2000. Solid waste management health and safety risks. Gainesville, FL: Florida Center for Solid and Hazardous Waste Management. Accessed August 30, 2009 at: http://www.hinkleycenter.com/publications/englereport..pdf. .

8. Rosenman, Kenneth D.; Reilly, Mary Jo; Schill, Donald P.; et al. 2003. Cleaning products and work-related asthma. Journal of Occupational and Environmental Medicine 45(5):556-563.

9. Smith, Gordon S. 2001. Public health approaches to occupational injury prevention: do they work? Injury Prevention 7:i3-10.

Reconciling Workplace Injury and Illness Data Sources

Nicole Nestoriak, Brooks Pierce, and John Ruser, US Bureau of Labor Statistics

Introduction

National estimates of nonfatal workplace injuries and illnesses are currently generated by the Bureau of Labor Statistics Survey of Occupational Injuries and Illnesses (BLS SOII), a comprehensive statistical program covering private industry and State and local government. The survey information is unique and of great value to the safety and health community in allocating prevention resources among several hundred diverse industries and occupations, across which workers' risks of injury and illness vary widely. For injuries and illnesses with days away from work, the survey also provides details that are critical to designing prevention strategies to protect workers. Survey data for SOII are provided by responding employers, who draw information from Occupational Safety and Health Administration (OSHA) logs and supplementary materials maintained by employers throughout the year. SOII is separate from other systems for recording workplace injuries and illnesses, including workers' compensation (WC), trauma registries and other administrative and survey data sources.

Two recent non-BLS research studies (Rosenman et al. (2006), Boden and Ozonoff (2008)) have raised the possibility that the SOII undercounts workplace injuries and illnesses that are within scope of the SOII. These studies are based on matching individual injury and illness cases in SOII to other data on workplace injury and illnesses cases, largely WC claims. The studies conclude that SOII and other data sources each miss injury and illness cases, leading to the conclusion that no single source of data can completely enumerate all cases.

While Rosenman et al. provided some evidence to explain differences in coverage of cases in the BLS and WC data, additional information about differences in the data in the two systems (and indeed other systems for capturing workplace injuries and illnesses) is necessary. To this end, BLS conducted research using a data file of matched SOII-WC data for Wisconsin created by Boden and Ozonoff. The research sought to identify factors that were associated with different levels of SOII capture rates, defined as the percentage of workers' compensation cases that were found in SOII[1].

Method

BLS obtained matched SOII-WC data for Wisconsin for 1998 to 2001, comprising approximately 217,000 distinct cases[2]. The file was created by Boden and Ozonoff by matching the lists of cases in the SOII with those in the WC administrative files.

The SOII is an annual establishment survey, currently with about 176,000 sampled private industry units nationally. BLS samples data at the establishment level rather than at the firm level. Firms with multiple sites or establishments may have some, none, or all of their establishments sampled in any given year. Data for a given survey year are reported to BLS in the first half of the year following the survey year. For more serious injury or illness cases involving at least one day away from work beyond the date of injury or onset of illness, the SOII collects detailed information describing the incident and the affected employee. Collected information includes the nature and source of the injury or illness, the part of body affected, the date of injury or illness onset, as well as the employee's name, date of birth, gender and race. These data elements, as well as information

---

[1] See Nestoriak and Pierce [2009] for more details about the study.
[2] All case totals in this report are weighted totals using SOII sampling weights.

on the employer, are used to help identify cases for the purposes of matching to WC administrative records.

The Wisconsin WC data are based on employers' first reports identifying the affected employee and circumstances of the case, augmented with supplemental reports documenting case duration and compensation payments. Data elements include worker-related fields such as employee name, date of birth, and gender; injury-related fields such as the date and duration of injury; and, employer-related fields such as company name and address, industry code, and a state employer identifier. The data include cases that were recognized after the end of the year of onset and some contested cases and negotiated settlements that are not separately identified.

Boden and Ozonoff linked WC claims to SOII cases using data elements common to both sources. The primary method of linking involved a deterministic match: a case was considered a match if both data sources listed the same values for eight or more key data fields related to the injury and the worker and company identities. Additional linked cases were identified using probabilistic record linkage techniques and human review. To maintain comparable scope in the two data sources, WC data are restricted to include cases at risk for SOII sampling. WC cases without lost work-time are excluded, as are cases occurring in companies not in the SOII sample.

The present study identified factors that were associated with different SOII capture rates, by means of tabulation and multiple regression.

Results

Whereas the SOII samples establishments, the WC data tend to reflect firm reporting. The WC data are insufficiently detailed to consistently determine whether an injury occurred at a particular physical location within the firm. This presents a data issue when a firm has multiple establishments, only some of which are sampled by SOII: is an injury case apparently missed by SOII truly a missed case, or rather is it an injury occurring at an establishment not in sample? While Boden and Ozonoff made a statistical correction for this, it is reasonable to assess the degree of matching separately for single and multiple establishment companies in Wisconsin.

Table 1. SOII Capture Propensity by Year of WC Filing

Year of WC Case Filing	Cases	Distribution	SOII Capture Propensity
Same Year as Survey Year	83,256	86.0	76.1
1 Year After Survey Year Close	12,406	12.8	48.0
2 Years After Survey Year Close	917	0.9	19.2
3 Years After Survey Year Close	203	0.2	4.9
4 Years After Survey Year Close	102	0.1	0.0
Total	96,884	100	71.8

Notes. Data are WC cases in Wisconsin single-establishment firms, 1998-2001.

SOII survey capture rates are higher when looking only at single-establishment firms. In particular, the SOII captures 77.5 percent of the estimated cases in this subset of the data. In multi unit and unknown status establishments, the capture rates are 62.2 and 52.8 percent respectively. The data where single-establishment status is unknown appears to behave more like the multi-establishment than the single-establishment subset of the data. The question that is not answered by these results is whether the lower capture rates for multi unit and unknown status establishments stem from the fact that these companies differ from single establishment companies or whether the statistical adjustment for matching survey to administrative census data was not entirely successful.

The timing of collection of injury and illness data differs between WC and SOII and may explain part of the discrepancy between the two data sources. In particular, the SOII collects data in the first six months of the year following the year of incidence and therefore will only contain cases that are recognized as a valid work-related case during or just after the survey year. Cases that are not recognized prior to data collection are not, obviously, included in the SOII counts. The WC administrative data, however, records cases up to two years following the date of incidence.

Table 1 shows case counts and the SOII capture propensity as a function of the WC system year of filing, for single establishment companies. A case with date of injury in 1998 and a WC system identifier indicating filing in 2000 would be included in the row "2 Years after Survey Year Close". Note that about 12.8 percent of cases are filed in the year following the survey year. A little over 1 percent of cases are filed with a greater lag. The final column shows the SOII capture propensity, defined here as the percent of WC cases that appear in the SOII.

Two broad facts are clear in these data. First, there are a substantial number of cases filed in WC after the close of the SOII survey year. Second, the SOII system capture propensity is much lower for cases filed in the WC system after the close of the SOII survey year. Together these facts suggest that the WC data include many cases that are not known to the SOII respondent, or are not resolved as work-related, at the time of survey response.

Aside from year of filing, we also know the order in which cases are entered into the WC system. Cases in the "1 year after" category disproportionately occur early in the filing sequence. About half of these cases appear to be filed early in the calendar year following the SOII survey year. For that half, the SOII capture rate is fairly high, approximately 60-65 percent. For the other half of the "1 year after" data, the SOII capture rate is approximately one-third. That is, the "1 year after" capture rate of 48.0 in table 1 masks variation within the year.

Both the SOII and WC databases contain information on the broad type of injury or illness, in the form of the "nature" of the case, which identifies the principal physical characteristics of the injury or illness. Nature of injury categories which one might reasonably view as severe, easily identifiable, or with sudden onset tend to be better captured by SOII in the single establishment matched data. For example, the capture propensities for amputations and severance cases are both about 90 percent. At least in these single establishment data, the vast majority of amputations are reported in SOII. Case types involving concussions, fractures, punctures and the like also tend to have relatively high SOII capture rates.

Case types such as lacerations, contusions, or strains, where one might expect somewhat greater heterogeneity of severity or ease of identification, tend to show average SOII capture rates, ranging from 71 to 75 percent of WC cases. These injuries are quite common, and documenting sources of heterogeneity within this subset of cases is a useful area for future work. Cumulative injuries such as inflammation or carpal tunnel are reported in the SOII much less frequently than the average case type, below 60 percent capture rate. These natures also tend to show longer than average lags between injury onset and WC filing. Presumably some of these cases develop too late for inclusion in the SOII data collection effort, or perhaps there is more ambiguity about whether such cases are work-related.

Discussion

Workplace injury and illness data sources differ in coverage for a variety of reasons. The SOII covers cases that meet OSHA criterion for recordability, while the WC data represent claims according to WC rules. Workers' compensation data are often limited to those cases lasting more than a certain number of days away from work (longer than a "waiting period"). Workers' compensation claims data may also include cases with only partial days away from work and may also include awards where the number of days away from work (if any) is uncertain. In contrast, the SOII case data include cases with at least one full day away from work. Thus, the scope of SOII and WC data may differ in terms of out-of-work duration and reconciling the two data sources must take this into account. SOII data are collected soon after the end of a reference year, while WC claims files remain open for several years. SOII data may not include cases that are recognized only after the reference year, or for which work-relationship is being contested at the time of the survey.

The results of further analysis of the Boden-Ozonoff matched SOII-WC data for Wisconsin suggest that legitimate differences among data systems and limitations of the undercount research methodology account for some of the discrepancy between the two data sets. For example, the research suggests there are some challenges in matching survey data to WC claims in the case of multi-establishment companies. The lower SOII capture rate for multiple establishment companies may reflect these challenges. However, there is also an indication that certain types of cases are less likely to be captured in the SOII, especially those that are more difficult to relate to the workplace and those with late onset or recognition. These latter cases may enter the WC data after the end of the survey year and may not be captured in the SOII.

The lessons of the research conducted thus far are that each data source alone may provide a useful but incomplete picture of workplace injuries and illnesses. In considering the usefulness of a data source for surveillance and prevention, it is important to recognize the factors that influence the capture of cases in that particular data source. WC data are limited by the legal requirements for filing a claim and by other restrictions on the cases that enter a WC data source, such as a minimum duration ("waiting period") before a case becomes compensable. National surveillance using WC data is limited by legal differences across States and differences in the scope of the data in the States. However, WC data have the strength of capturing cases that may not be recognized until after (sometimes long after) the end of a particular reference year.

There is need for more research to determine the factors that might account for differences between SOII, workers' compensation and other workplace injury and illness data sources. To this end, BLS continues to conduct intramural research matching SOII to WC data in additional states. BLS is also supporting extramural research for a small number of States to conduct additional studies: 1) multiple data source enumeration of amputations and carpal tunnel syndrome cases and 2) WC-SOII matches followed by employer interviews regarding OSHA recordkeeping and WC claims reporting practices.

References

Boden L, Ozonoff A [2008] Capture-Recapture Estimates of Nonfatal Workplace Injuries and Illnesses, Annals Epidem. 18(6): 500–06.

Nestoriak N, Pierce B [2009] Comparing Workers' Compensation claims with establishments' responses to the SOII, Monthly Labor Rev. 132(5): 57-64.

Rosenman K, Kalush A, Reilly M, Gardiner J, Reeves M, Luo Z [2006] How Much Work-Related Injury and Illness is Missed by the Current National Surveillance System? J Occup Environ Med. 48(4): 357–65

Methodological Challenges in the Liberty Mutual Workplace Safety Index:

Working Towards a Future Model

Helen Marucci-Wellman, ScD, Liberty Mutual Research Institute for Safety

Introduction

Although occupational injuries are among the leading causes of death and disability around the world, the burden due to occupational injuries has historically been under recognized, resulting in a missed opportunity to address a priority public health problem.

The Liberty Mutual Workplace Safety Index (LMWSI) was developed in 2000 to provide a national estimate of the annual workers' compensation cost burden of the most disabling injuries and illnesses in the U.S. For the purpose of benchmarking the most disabling workplace injuries, our approach is predicated on the assumption that workers' compensation costs are a reasonable surrogate measure of injury severity, incorporating, as they do, wage compensation benefits paid for duration away from work and the total cost of medical treatments for those injuries reported. Using such a metric, we report on the leading causes of work-related injuries and illnesses in the U.S. The purpose of this paper is to introduce the LMWSI and outline technical challenges.

Methods

LMWSI annual estimates are derived using three national data sources. For injuries and illnesses resulting in more than 5 lost work days, the mean costs of Liberty Mutual (LM) workers' compensation claims are multiplied by Bureau of Labor Statistics (BLS) frequencies for each two-digit BLS Occupational Injury and Illness Classification System (OIICS) event category. The top 10 disabling event categories are determined from the rank of the fractional component cost of each event group to the total. Finally, the National Academy of Social Insurance (NASI) estimate of total benefits paid is used to adjust estimates of the projected national burden of each category (see Figure 1).

The Three Components of the Index

A) Liberty Mutual Mean Cost

All claims incidents for each year are extracted, including the Liberty assigned cause-and-nature of injury codes and narrative text information on the event sequence involved in the injury. Each case is then coded for event, or "Cause of Injury," according to the BLS Standardized Occupational Injury and Illness Classification protocol using a combination of automated and manual coding methods. The coding methods include: 1) a crosswalk from Liberty Mutual's "cause codes," 2) automatic narrative text filtering and 3) manual assignment by a coding panel. Mean cost estimates are calculated using the paid costs of closed claims. For open claims an estimate is developed for the ultimate cost of the claim. Each year all estimates are discounted to 1998 dollars. LM mean costs are adjusted prior to calculating total costs incurred for each claim and both the real and nominal values are used to project costs for the final index. Indemnity payments are adjusted using the Consumer Price Index (CPI) and the medical payments are adjusted using the Medical Services Price Index (MSPI)

An obvious issue in using the data from a single carrier is that the sample of industries is not likely representative of the nation as a whole. We sought to accommodate this by using BLS estimates to determine the relative frequency distribution component of national cost by event category.

B) Bureau of Labor Statistics Frequency Estimate

The BLS Survey of Occupational Injuries and

Figure 1. Methodology

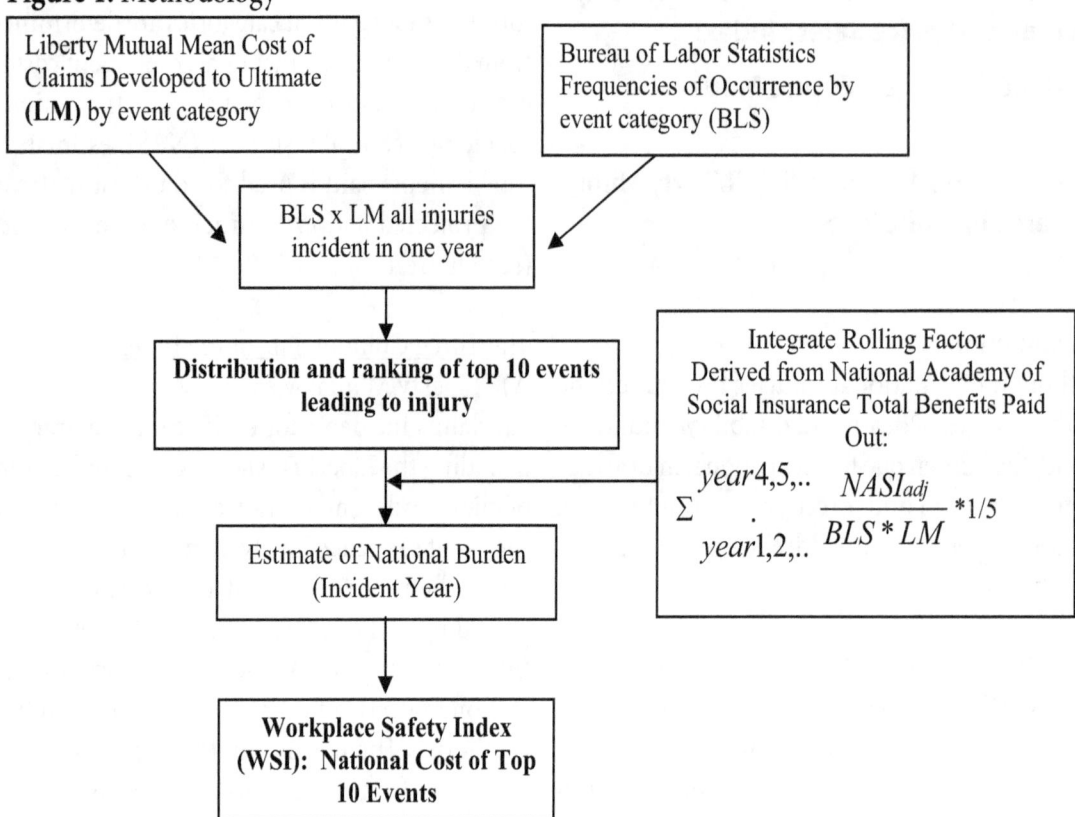

Illnesses (SOII) provides annual estimates of the number and incidence rates of occupational injuries and illnesses among private sector workers at the National level. Estimates are based on a sample of illness and injury logs maintained by private sector employers as required under OSHA record-keeping guidelines (OSHA logs). Excluded from the survey are self-employed workers, farms with fewer than 11 workers, and employees of federal, state and local government agencies and certain other sectors. Because of these exclusions BLS estimates will undercount the true national frequency of occupational injuries. Recent studies have also demonstrated an undercount for varied reasons (Leigh et al., 2004; Rosenman et al., 2006; Boden and Ozonoff, 2008).

Additionally, while both of these data sources have limitations, we assume the relative rankings by event category for both data sources to be robust and representative. For example, it is assumed that the proportionate frequency distribution by event category is uniformly distributed even though the BLS data are an undercount of the total frequency of injuries. This assumption needs to be further investigated. While it is very possible that there are strong biases in the distribution of cost by event category due to reporting biases by event group (in both the BLS survey and workers' compensation), we attempt to minimize this bias by including only the most severe injuries (e.g. cases losing more than 5 days of work).

C) National Academy of Social Insurance (NASI)

The National Academy of Social Insurance each year publishes estimates of the total benefits paid out in workers' compensation in the U.S. In order to adjust for the potential underestimate of using the BLS x LM as an estimate of total national cost burden, it is adjusted using the annual NASI estimate of all paid benefits. Specifically, we apply a rolling factor to the BLS x LM direct cost burden. This factor is essentially a rolling five-year average of the ratio between the NASI estimate benefits paid for claims with days away from work each year and LMWSI derived national estimate (BLS x LM). The NASI estimate is a calendar year estimate while the LMWSI-derived estimate is an incident year estimate, but equilibrium or a steady state should cause these two estimates to converge (NASI, 2005). We also note that the NASI estimate includes benefits paid to individuals excluded in the BLS survey.

Summary Findings

The overall LMWSI values from 1998 through 2006 are shown in Figure 2.

The estimated national cost burden for claims incident in 2006 was $48.6 billion or $37.2 billion expressed in 1998 dollars. This was the fourth year that there was a decline in the inflation-adjusted figures; however, the decline showed a slowing from the prior two years (see Figure 3). Since 2002, the peak year of the Index, there has been a large overall decline in real terms ($45.6 billion to $37.2 billion in 1998 dollars).

The fractional component cost and final burden estimate, both nominal and real (adjusted for inflation from 1998) of each event group, is shown for the top 10 ranked categories in Table 1.

Creating An Index to Monitor Trends

An indexing of values allows us to understand the possible trending components of each category more thoroughly. If all values are indexed to 1998 (in real terms) the trending of the index is sensitive to changes in either of the two factors (frequency and cost). Two examples of such sensitivity analyses are provided in Figures 4 and 5 for the categories of Overexertion and Fall to Lower Level.

From the sensitivity analysis we are able to see the trends in the two separate data sources. The upper curve illustrates the percent increase from 1998 in cost if the frequency were held constant for all 9 years (and at 1998 values). Alternatively, if cost were held constant for all years the percent decline from 1998 values in the frequency is shown in the bottom curve.

Figure 2. LMWSI 1998-2006 (1998 Dollars)

The reduction in the overexertion category to below 1998 values in real terms is probably not associated with a reduction in the workforce between 1998 and 2006 because there is a similar steep downward trend in the BLS rates for this category (53.8 to 30.8 per 10,000 workers 1998 and 2006 respectively, BLS, 2000 & 2008.) However there are many other possible reasons, for example, a reduction in risk due to fewer workers in environments with high risk for overexertion or a disproportionate reduction in workers exposed or workers' willingness to report. It may also be due to advances in the research and prevention of these disabling disorders. Less costly medical procedures, lower average severity or improvements in return to work strategies could clearly reduce costs further for this category. What is certain is the Index demonstrates that the burden to workers' compensation systems for injuries caused by overexertion events is falling. Further, while the frequency of overexertion cases has dropped steadily in the last nine years, costs increased steadily through 2003 but have recently leveled off. Figure 4 shows that a 30 point drop in frequency (keeping cost constant) has a larger effect on the trending of the overexertion category than a 40 point increase in cost.

The analysis of fall to lower level category is also shown (Figure 5) because the components trend differently. For disabling falls to the lower level, it appears that little has changed in prevention or cost reduction once an injury occurs. Index values are almost 20 points higher in real terms than in 1998 and this category has also moved from 5th to 3rd rank.

Discussion

The total national direct, insured burden due to the most disabling injuries incident in 2006 is estimated to be 48 billion. We should recognize that this "burden" is limited to only a small portion of the entire burden to society of workplace injuries and illnesses. The costs incorporated into this

Figure 3. Annual Real Growth Rates LMWSI 1998-2006

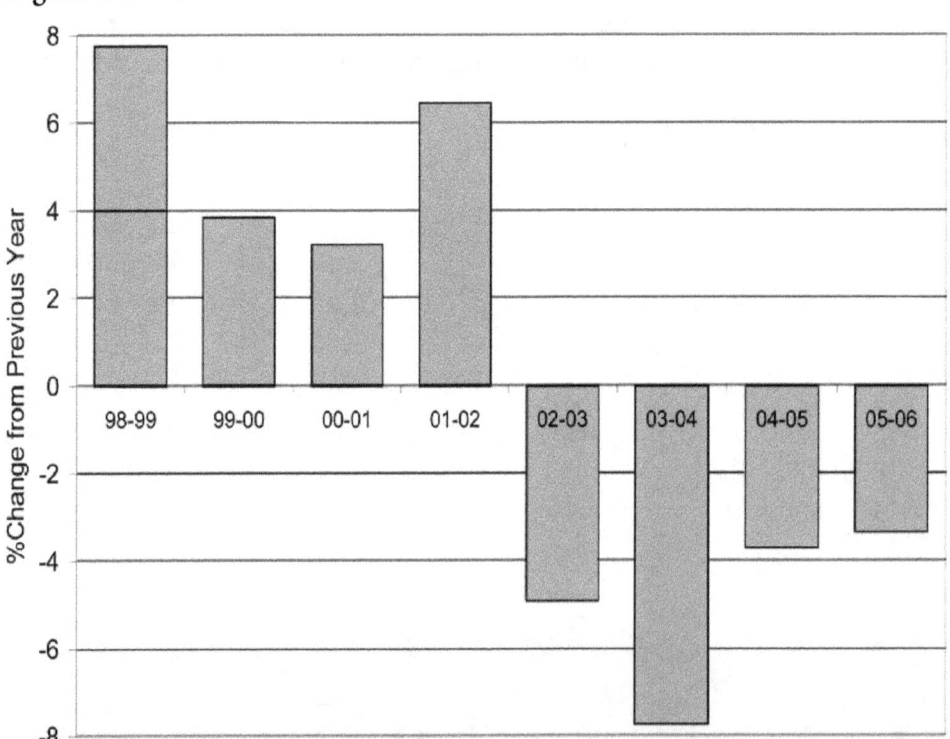

Table 1. Disabling Events Ranked by Percentage of Workers' Compensation Benefits and Total Costs of Greater than 5 Lost Workday cases in the United States 2006

Rank	Event Description	Proportion, 2006 Benefits	Est. Cost, 2006 dollars[5]	Est. Cost, 1998 dollars[6]
	Total	100	$48,644,606,743	$37,238,141,665
1	Overexertion	25.7	$12,448,323,690	$9,566,023,909
2	Fall on same level	13.3	$6,446,795,936	$4,943,666,665
3	Fall to lower level	10.8	$5,265,854,229	$4,015,543,705
4	Bodily reaction	10.0	$4,834,090,520	$3,710,563,272
5	Struck by object[1]	8.9	$4,325,992,583	$3,308,991,612
6	Struck against object[2]	5.1	$2,501,091,960	$1,909,732,734
7	Highway incident	4.9	$2,378,428,567	$1,812,289,358
8	Caught in or comp[3]	4.4	$2,141,167,273	$1,632,551,968
9	Repetitive motion	4.0	$1,955,814,003	$1,504,042,237
10	Assaults[4]	0.9	$442,946,461	$340,658,339
	All Other	12.1	$5,904,101,520	$4,494,077,866

[1] Struck by object or equipment
[2] Struck against object or equipment
[3] Caught in or compressed by equipment or objects
[4] Assaults and violent acts by person(s)
[5] Estimated total workers' compensation actual costs for U.S. private industry
[6] Estimated total workers' compensation costs for U.S. private industry in 1998 dollar

Table 2. Estimated Real and Nominal Growth in Workers' Compensation Total Costs for Events for Cases with Greater than 5 Lost Workdays, United States 1998-2006

		Estimated Growth in Costs	
Rank	Event Description	Nominal, 1998-2006	Real, 1998-2006[5]
	Total	30.9	0.4
1	Overexertion	35.0	-4.8
2	Fall on same level	59.3	17.9
3	Fall to lower level	36.2	17.9
4	Bodily reaction	35.9	8.5
5	Struck by object[1]	32.5	7.3
6	Struck against[2]	37.0	16.2
7	Highway accident	5.8	-4.4
8	Caught in or comp[3]	18.1	12.5
9	Repetitive motion	14.9	-35.3
10	Assaults[4]	8.7	-15.7
	All Other		

[1] Struck by object or equipment
[2] Struck against object or equipment
[3] Caught in or compressed by equipment or objects
[4] Assaults and violent acts by person(s)
[5] Adjusted for inflation from 1998: Indemnity and expense costs adjusted using the Consumer price index (CPI), Medical costs adjusted using the Medical Services Price index (MSPI)

Figure 4. Overexertion: Index Sensitivity Analysis 1998-2006

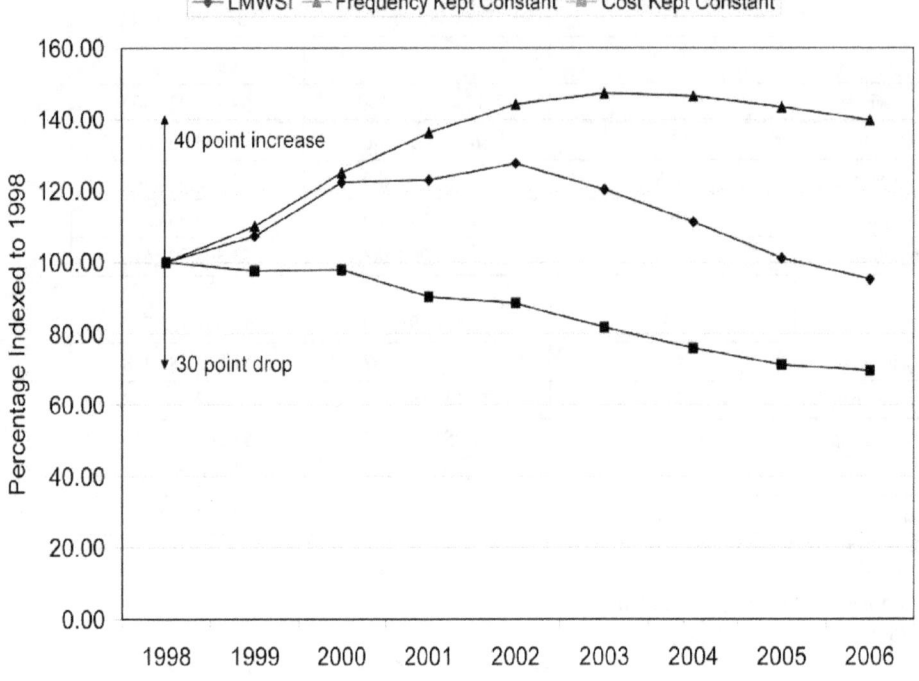

Figure 5. Fall to Lower Level: Index Sensitivity Analysis 1998-2006

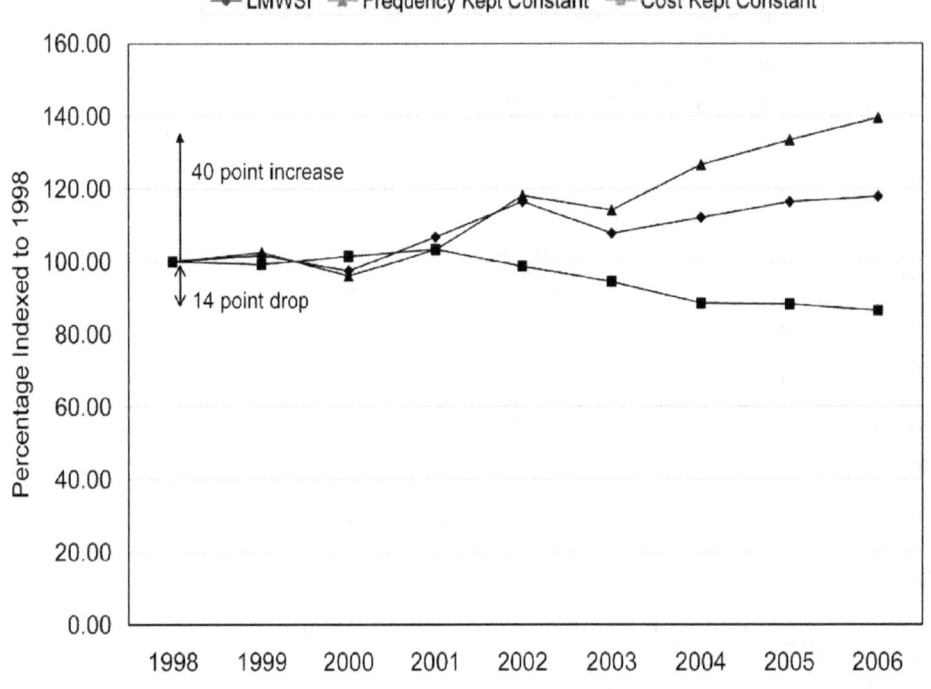

Index include only direct costs of benefits paid out to individuals for only the most disabling injuries reported and accepted into the workers' compensation system. It follows that some workplace injuries or illnesses not reported in workers' compensation are not included. The workers' compensation system also does not provide information about wages lost during waiting periods and reimburses 66% or less of workers' wages in most jurisdictions once compensation begins. This analysis included neither the medical costs to workers missing between 1 and 4 days nor actual replacement costs for all waiting periods. During development of the Index, this value was compared with other estimates of national burden (NSC, 2000; Leigh, 1997 & 2000), and found to be representative of the burden comprised from WC benefits as defined.

We understood during the development of the Index that there were several limitations resulting from combining data from three different national datasets. From its inception, we have considered that future analyses would lead to the introduction of modifications to our methods and that, in order to maintain an acceptable level of inter-year comparability, these introductions would need to be carefully managed. Introducing improvements at specific points in time compromises analyses of historical trends since the data are no longer comparable.

Of most significance is the determination of whether the biases introduced by any of the data sources are differential with regards to the proportionate magnitude of cost by event category or the way costs change annually by category. A differential effect would limit the value of the highlighted contributions of each cause to total burden as well as the interpretation of year to year growth in the index. We have begun to investigate this by sensitivity analysis and will continue to develop these methods.

For both the Liberty Mutual and Bureau of Labor Statistics components it is critical that the distribution of the component burden (by mean cost and frequency, respectively) is representative of the nation. If we find differential biases by event category, we will need to develop new models to adjust for this bias.

The accuracy of LM mean cost data as an estimate of the "true" WC cost burden by event group may be compromised by a) the LM "book of business," b) the influence of "extreme values," and c) biases in compatible case event classifications. Therefore, one such modification could be to use an estimate of market share to weight the data to provide a more representative set of mean costs and representative set of high cost claims (outliers) which will drive the mean cost of any one category. Methods for better, more reliable classifications should also be implemented as computer classification approaches become available. When combining the Liberty Mutual data with Bureau of Labor Statistics data on event category, it is critical that compatible case definitions are used, which requires matching of claims data to BLS defined characteristics. This can be quite challenging with the limited information in claims narratives and tens of thousands of claims to classify each year.

Of concern, also, is the undercount in BLS frequencies due to the exclusion of important risk groups from the survey and the sample design. Potentially, this data might be accommodated by added weightings for undercounts for some industries or for small employers.

Finally, integrating the NASI cost estimate might be improved by examining any distributional differences in exposure and also whether the proportion of total losses paid for indemnity cases is consistent between LM and NASI. In addition to this, we should investigate the best way to apply this factor to understand the true trend in incident costs.

Acknowledgments:
Recent releases of the Index have involved Helen Marucci-Wellman, Helen Corns, Simon Matz, and Ted Courtney. The LMWSI was created between July and November 2000 at the Liberty Mutual Research Institute for Safety. Then Director Tom Leamon envisioned and created a team of research scientists to develop the index including: Ted Courtney, Helen Marucci-Wellman, Gary Sorock, Barbara Webster, Simon Matz and Radek Wasiak. Current Institute Director Ian Noy, and former Senior Vice President for Loss Prevention Karl Jacobson, also provided executive oversight during the first decade of the Index. Additionally, we are indebted to many in the workers' compensation research community who have contributed their thoughts, reviews, or comments to the Index program at various stages.

References
Bureau of Labor Statistics (BLS). [April, 2000] Case and Demographic Characteristics for Work-related Injuries and Illnesses Involving Days Away From Work: Supplemental Tables for 1998. http://www.bls.gov/iif/oshcdnew.htm.

Bureau of Labor Statistics (BLS). [April, 2008] Case and Demographic Characteristics for Work-related Injuries and Illnesses Involving Days Away From Work: Supplemental Tables for 2006. http://www.bls.gov/iif/oshcdnew.htm.

Boden LI, Ozonoff AL. [2008] Capture-recapture estimates of nonfatal workplace injuries and illnesses. Epidemiol. 18(6):500-6.

Leigh JP, Markowitz SB, Fahs M, Shin C, Landrigan PJ. [1997] Occupational injury and illness in the United States. Estimates of costs, morbidity, and mortality. Arch Intern Med. 157(14):1557-68.

Leigh JP, Markowitz SB, Fahs M, Landrigan PJ. [2000] Costs of Occupational Injuries and Illnesses. University of Michigan Press

Leigh JP, Marcin JP, Miller TR. [2004] An estimate of the U.S. Government's undercount of nonfatal occupational injuries. J Occup Environ Med. 46(1):10-8.

National Academy of Social Insurance (NASI). [July, 2005] Workers' Compensation: Benefits, Coverage, and Costs, 2003. pp 38-40.

National Safety Council (NSC). [2000] Injury Facts, 1999 Edition. Itasca, IL: Author.

Rosenman KD, Kalush A, Reilly MJ, Gardiner JC, Reeves M, Luo Z. [2006] How much work-related injury and illness is missed by the current national surveillance system? J Occup Environ Med. 48(4):357-65.

Harmonizing Existing Databases Counting Workplace Injuries and Illnesses

Arthur Oleinick, MD, JD, MPH, University of Michigan
Brian Zaidman, B.A., Minnesota Department of Labor & Industry

Introduction

The Bureau of Labor Statistics annual Survey of Occupational Injuries and Illness (BLS SOII) (http://www.bls.gov/iif) has provided nationwide estimates of work injuries and illnesses since 1972, a little more than one year after the passage of the Occupational Safety and Health Act.[1] Initially, survey data were supplemented with injury characteristic data for cases with days away from work (DAFW) from state workers' compensation (WC) systems through the Supplementary Data System in which some 36 states ultimately participated. However, in 1992, the SOII was revised so that injury characteristic information for cases with at least one DAFW was obtained directly from the survey. In addition to BLS, the National Center for Health Statistics has periodically included questions on workplace injuries and illnesses in its National Health Interview Survey (NHIS) (http://www.cdc.gov/nchs/nhis.htm) and the Consumer Product Safety Commission's National Electronic Injury Surveillance records injury origin for injuries seen in hospital emergency departments (ED) (the work injury/illness data are available at http://www2a.cdc.gov/risqs).

Two recent studies using capture-recapture methodology to compare case-specific BLS and WC data in the period 1998-2001 reported undercounts of DAFW cases in the BLS system of 25-68% and 5-35% in the WC state systems studied.[1-2] An earlier study by the present authors[3] comparing total counts for comparable DAFW groups in the Minnesota BLS SOII and WC databases in the period 1992-2000 found that the BLS SOII had 88-93% of the number of WC cases recorded. Smith, using estimated counts from NHIS data for

Table 1. Scenarios: Eligibility for MN WC wage replacement and count as BLS DAFW injury.

Scenarios	1	2	3	4
Mon	Inj. & Leaves[1]	Inj. & Stays[2]	Inj. & Stays	DAFW[3]
Tues	DAFW	DAFW	Works	Returns
Wed	DAFW	DAFW	DAFW	
Thurs	DAFW	DAFW	DAFW	
Fri	Returns	Returns	Returns	Inj. & Leaves
BLS DAFW	3	3	2	1
Eligible	Y	N	N	Y
Days Paid	1	0	0	1

1 – Leaves work before end of shift on day of injury.
2 – Stays at work until end of shift on day of injury.
3 – Day away from work following injury on previous Friday.

1997-1999, reported that BLS underestimated the number of DAFW injuries by 30%.[4] The National Institute for Occupational Safety and Health, extrapolating from ED visits for work injuries and illnesses regardless of DAFW, estimated a total count of work injuries in 1998 substantially exceeding that reported by BLS for the year.[5]

Because of the focus of this session, this summary presentation is confined to a consideration of the methodologic issues raised by comparisons of BLS DAFW and WC wage compensation case databases.

The information provided in this summary presentation is based upon data and information in Oleinick and Zaidman (2009)[6] and is authorized under the copyright agreement.

Methods

The basic methodologic problem in comparing case-specific BLS DAFW cases with WC cases eligible for wage compensation in the capture-recapture studies and our past and current study is that both case groups are differently defined subsets of the larger populations of work injuries and illnesses covered by the two systems. We regard Minnesota data, included by one of the capture-recapture studies,[7] as an appropriate prototype for considering these methodologic issues based on our review of various statutory/regulatory provisions and practices in several other states used in the capture-recapture studies. To create comparable subsets in the two databases, we restricted the BLS DAFW case count to those cases with ≥ 4 DAFW and the WC wage compensation cases to those cases paid wage compensation for ≥ 1 day temporary total disability (TTD) with ≥ 4 DAFW (wage compensation payments begin after 3 days of disability and are not retroactive to the first day of disability until the worker has been disabled for 10 days). The methodologic problem resolved by our approach is illustrated in Table 1.

Table 2. Concordance (%): ≥ 4 DAFW BLS SOII Count and TTD/PTD MN Indemnity Claims, 1998 – 2001.

Count of BLS DAFW, Private Industry and State/Local Government	149,442
Upper Estimated Count of ≥ 4 BLS DAFW	83,753
Lower Estimated Count of ≥ 4 BLS DAFW	78,089
MN TTD/PTD Indemnity Claims Paid[1]	108,426
MN Indemnity Claims Paid for ≥ 4 DAFW	*93,013*[2]
Upper Estimate of Concordance between BLS DAFW and MN TTD Claims	*90.04%*[2,3]
Lower Estimate of Concordance between BLS DAFW and MN TTD Claims	*83.95%*[2,3]

1 – MN Indemnity claims filed by April 1 of year after injury year and payments recorded through mid-March 2002.
2 – Italics indicate counts from 1998 – 2000 and estimates for 2001 based on counts from 1992 – 2000.
3 – If only private industry is included, the upper and lower concordance estimates are 89.23% and 83.50%, respectively.

Although the date of injury is the same for the first three scenarios, the pattern of time away from work differs in each case. Minnesota data indicate that about one-third of workers follow each of the patterns indicated, generally without regard to day of injury. However, although all three scenarios would qualify for inclusion in the BLS DAFW count, only the first would receive wage compensation because the three-day work disability count begins with the first day of lost time (FDLT). Cases not paid wage compensation would, however, have their medical costs covered.

The fourth example indicates that an injured worker can qualify for wage compensation with as little as one DAFW because of the counting algorithm. In contrast, the way the counting algorithm works means that BLS SOII cases with >3 DAFW would necessarily qualify for wage compensation. Since the BLS SOII does not collect information on the FDLT, there does not appear to be a software solution that would identify WC comparable cases among those injured workers with ≤ 3 DAFW in the BLS SOII.

Our work is based on the premise that verifying completeness of ascertainment in the two subgroups we compare is an important effort by itself because they contain most of the information on work disability producing time away from work. Moreover close agreement in these two subgroups would suggest, but not prove, that other subgroups are reasonably ascertained by the two databases.

In addition to considering legal criteria as a potential confounder in studies comparing BLS and WC cases, we obtained information of the percent completion on data elements in the two systems that were used to match cases in the capture-recapture studies to estimate the potential loss in matching on this basis.

Results

Table 2 summarizes the percent concordance in counts between comparable groups in the Minnesota BLS SOII and WC databases. Both comparison groups have at least 4 DAFW. BLS SOII did not collect information on the FDLT on the survey form (Form N) used in Minnesota during the study period nor did the form itself specify whether the worker qualified for wage compensation.

The comparison WC group consists of those injured/ill workers who received any number of days of temporary total disability/permanent total disability (TTD/PTD) and had ≥ 4 DAFW. Workers receiving only temporary partial disability (TPD- "light duty", partial work days) or

Table 3. Capture – recapture Undercount Estimation: MN WC indemnity claims and BLS SOII, 1998 – 2001.

	MN WC Indemnity Eligible		
BLS ≥1 Day DAFW	No Report	Report	Total
No Report	a[1]	b	a + b
Report	c	d	c + d
Total	a + c	b + d	a + b + c + d

1 – Estimated by capture – recapture, Boden and Ozonoff (2008)

permanent partial disability (PPD- according to degree of impairment, set out in a statutory schedule) but without payments for full days away from work and workers who received a lump sum payment for any medical and/or disability costs as a result of a stipulation or settlement agreement were excluded from the comparison. There is simply no way to determine from the forms submitted to the state whether any of the injured/ill workers in the last three groups had DAFW in addition to the type of work disability for which they were paid.

The BLS SOII count range of 78,089-83,753 is 84-90% of the comparable cases found in the WC system, depending on the assumptions used to estimate the BLS SOII cases with 3 DAFW in the 3-5 DAFW group. This range is little changed when the comparison is restricted to private industry. We concluded that a minimum estimate for undercount in the BLS SOII survey was in the range 10-16%. Inclusion of the one-third of stipulation/settlement cases who recorded a FDLT on the WC first report of injury could shift this range lower by as much as 2%. Inclusion of later-filed WC cases with wage compensation eligibility would further lower the range by about 1%.

Table 3 is adapted from the report by Boden and Ozonoff (2008)[3] to emphasize its underlying structure as a standard 2X2 table. They used adjusted BLS sampling weights and assumed that injury risk was homogeneous throughout a firm containing a sampled establishment to obtain statewide estimates of counts from the samples of BLS and WC cases compared. Their estimated Report column total of 112,251 for WC wage compensation cases with reports (b+d) serves as the denominator for the BLS SOII undercount estimate (substituting marginal totals yields the same estimate) but appears too high at 32.4% in light of the count of such cases reported to Minnesota for the study period. Excluding the estimated state/local government cases and private industries' cases excluded by Boden and Ozonoff from the actual count of WC cases paid, the number of private industry cases paid either TTD/PTD or closed by stipulation/settlement is approximately 95,000 for the period. This total includes only those stipulated/settled cases with an entry for the FDLT suggesting at least one day of acute work disability. In contrast, their column total of 112,251, (b+d), suggests that they decided to include the cases paid only TPD and/or PPD and stipulated/settled cases. Since the estimated numerator cell with 36,355 unmatched cases (b), is likely to be disproportionately affected by these inclusions, we believe that their BLS undercount percent would be smaller, particularly in light of the effect of missing data in the BLS and WC databases.

Similarly, we believe that the Boden and Ozonoff (2008)[3] estimate for the WC undercount is too high at 35.2%. We think this is largely attributable to the lack of information regarding a FDLT so that their estimate of 41,238 (c) as the numerator of the WC undercount rate, would include a large number of cases with ≤3 DAFW ineligible for wage compensation because of the eligibility counting algorithm or the absence of a required data element in the BLS SOII survey. Alternatively, if an arbitrary threshold number of DAFW such as three were used to determine wage compensation eligibility for this cell alone, we believe it would call into question their use of BLS sampling weights to extrapolate state totals.

Conclusions

Given the wide variation in undercount percents resulting from the use of various methodologies and databases, more precisely targeted information is needed before any redesign of the current BLS SOII survey is undertaken.

References

1. Abraham KG, Weber WL, Personick ME. 1996. Improvements in the BLS safety and health statistical system. Monthly Labor Review 119(4):3-12.

2. Rosenman KD, Kalush A, Reilly MJ, Gardiner JC, Reeves M, Luo Z. 2006. How much work-related injury and illness is missed by the current national surveillance system? J Occ. Environ. Med. 48(4):357-365.

3. Oleinick A, Zaidman B. 2004. Methodologic issues in the use of workers' compensation databases for the study of work injuries with days away from work. I. Sensitivity of case ascertainment. Am J Ind Med. 45:260-274.

4. Smith GS, Wellman HM, Sorock GS, Warner M, Courtney TK, Pransky GS, Fingerhut LA. 2005. Injuries at work in the US adult population. Contributions to the total injury burden. Am J Pub Hlth. 95(7):1213-1219.

5. Division of Safety Research. National Institute of Occupational Safety and Health. 1998. Surveillance for nonfatal occupational injuries treated in hospital emergency departments-- United States 1996. MMWR- Morbidity and Mortality Weekly Report. 47(15):302-306.

6. Oleinick A, Zaidman B. 2009. The law and incomplete database information as confounders in epidemiologic research on occupational injuries and illnesses. Am. J. Ind. Med. In press.

7. Boden LI, Ozonoff AL. 2008. Capture-recapture estimates of nonfatal workplace injuries and illnesses. Ann. Epid. 18(6):500-506.

The views expressed in this paper are those of the authors and do not reflect official policy or position of the Department of Labor and Industry, State of Minnesota.

Overview of an Insurance Carrier's Service Sector Data

Adam L. Seidner MD MPH, Senior Attending, Middlesex Hospital

Introduction

One of the strategic goals for the National Occupational Research Agenda Services Sector is to reduce the incidence and severity of occupational disease and disability for higher risk populations of services workers and for those employed in emerging technologies. There is a continuum from data, to information, to knowledge. New ways of collecting, looking at, and analyzing data are important. The need exists for more and better health indicators and new classes of information to enhance understanding of relationships in workers' compensation.

Reviewing an Insurance Carrier's Service Sector Data can help us to understand the strengths and limitations of its workers' compensation database. Analysis of insurance databases may lead to a better understanding of the occupational risks and interventions needed. It may be useful to consider insurance databases as part of surveillance information systems for illnesses, injuries and exposures in higher risk service industries and occupations. Understanding how an insurance database is populated and the elements involved allows for its proper use and interpretation.

Background

Insurance companies have multiple databases. Databases include client and policy information, billed invoices, and claim data. One insurance carrier developed a Managed Care Database by combining billing and claim data. An individual claim will have many bill lines. Analysis by claim is necessary to understand how the claim developed and identify trends.

Data Collection Methods

A sample of data from a Managed Care Database (MCDB) was selected based on a specified date range, job class, data fields, and state claim frequency. The data fields for the sample database are listed in Table 1.

Analytical Data Quality Review and Overview

Additional information needs include expanding the availability, accessibility, and quality of Workers' Compensation data. Holes in information do exist within the data; and barriers to overcoming missing information.

This project utilized a Workers' Compensation data set to determine the type and scope of data.

Missing data is something that can impact interpretation. Five claims did not have a state noted. State was assigned based on the benefit state. The benefit state may be different than the state where the workers' compensation claim occurred, where the company is based, or where the claimant lives. The majority of claims came from California, New York, Texas, Virginia followed by Washington DC, Maryland, Mississippi, Minnesota, Delaware, and South Dakota. Additional benefit states were in the database (Table 2) and these resulted from the worker moving their residence and care or, in some cases, the worker having a choice of which state to declare they receive benefits.

Missing codes were different for National Council on Compensation Insurance (NCCI) codes compared to ICD-9 codes (Table 2). The workers' compensation insurance industry works with the NCCI type of coding system. There was a significant number of missing ICD-9 codes (624) compared to only 6 from the NCCI field.

Table 1. Data Fields

CLAIM NUMBER	INJURY TYPE
POLICY NUMBER	EMPLOY DATE YYMD
POLICY EFFECTIVE DT YYMD	EMPLOY DATE MDYY
POLICY EFFECTIVE DT MDYY	OCCUPATION
BENEFIT STATE	ATTORNEY INVOLVED IND
ACCIDENT DATE YYMD	DISABILITY BEGIN DT YYMD
ACCIDENT DATE MDYY	DISABILITY BEGIN DT MDYY
AGE	RET WORK DATE YYMD
GENDER	RET WORK DATE MDYY
AREA FIELD OFFICE	RTN WORK DATE YYMD
FILE PREFIX	RTN WORK DATE MDYY
LOSS DATE YYMD	AVERAGE WEEKLY WAGE
LOSS DATE MDYY	BILL CONTROL NUMBER
TOTAL CLAIM RESERVE	DIAG 1
TOTAL MEDICAL RESERVE	DESC 1
TOTAL EXPENSE RESERVE	TOTAL INVOICE AMT
AMOUNT OF CLAIM PAID	TOTAL PAID AMT
AMOUNT OF MEDICAL PAID	PAY RELEASE DT YYMD
AMOUNT OF EXPENSE PAID	PAY RELEASE DT MDYY
NCCI INJURY CODE	CLAIM FILEASOF
NCCI BODYPART	PROVIDER NAME
NCCI NATURE OF INJURY	SERVICE FROM DT YYMD
NCCI CAUSE OF INJURY	SERVICE FROM DT MDYY
REFINAL DATE YYMD	SERVICE THRU DT YYMD
REFINAL DATE MDYY	SERVICE THRU DT MDYY
CLOSE DATE YYMD	BILL TYPE
CLOSE DATE MDYY	PAY ISSUE DT YYMD
STATUS	DATE PAYMENT ISSUED MDYY
SIC CODE	JOB CLASS

Table 2. Select Missing Data (number of claims)

CM	0
CB	0
State	5
Gender	5
NCCI	14
Age	16
Job Class	40 (CB – 5, CM-35)
NCCI Diagnosis	6 (CB – 5, CM- 1)
ICD-9 Diagnosis	624 (CB – 145, CM- 479)

Table 3. Descriptive Analysis of Claims Services Sector Data

Date Range: January 1, 2006 – January 30, 2008
Number of Bill Lines: 15,613
Number of Claims: 1,976
Injured Worker's Age Mean: 37.58 years (SD 13.49)
Gender Distribution: Female = 793   Male = 1183
Claim Type: Claims Medical (CM): = 1154 [58.4%]   Claims Both (CB) = 822
Open vs. Closed Distribution: Open = 1850   Closed = 126

Table 4. Top 14 Job Classes and Diagnoses

Job Class	Job Title	Frequency	ICD-9 code	Description
9050	Room Attendant	166	847.2	Sprain Lumbar Region
8391	Technician	165	847	Sprain Of Neck
8810	Customer Service Rep	111	840.9	Sprain Shoulder/Arm NOS
4304	Pressman	108	845	Sprain Of Ankle NOS
9155	Film Crew	100	724.2	Lumbago
9008	Janitor	75	722.1	Lumbar Disc Displacement
9061	Housekeeping	53	846	Sprain Lumbosacral
9014	Cleaner	51	883	Open Wound Of Finger
9079	Dishwasher	45	719.46	Joint Pain-L/Leg
9154	Stagehand	44	844.9	Sprain Of Knee & Leg NOS
8044	Product Tech	42	724.4	Lumbosacral Neuritis NOS
8742	Dist Sales Mgr	37	959.9	Injury-Site NOS
8293	Warehouse Worker II	36	719.41	Joint Pain-Shoulder
9053	Massage Therapist	34	354	Carpal Tunnel Syndrome

Table 5. Most Frequent NCCI Causes of Injury

NCCI Cause Of Injury	Frequency
56-Strain From Lifting	131
31-Fall Or Slip Not Classified	90
33-Fall On Stairs	76
32-Fall On Ice Or Snow	55
29-Fall On Same Level	51
81-Struck By Not Classified	29
98-Cumulative Not Classified	17
99-Missing	14
97-Strain From Repetitive Motion	13
17-Object Lifted Or Handled	11
60-Strain Not Classified	11
70-Strike/Step On Not Classified	10

Analysis

There were 1,976 established claims between January 1, 2006 and January 30, 2008 (Table 3). They had 15,613 bills and the majority of claims was still active and open (1,850). Claims are designated open or closed based on whether payments or medical care continues to occur. Claims are further divided into Claims Medical (CM) and Claims Both (CB). These are important distinctions. A claim that has no lost time from work and only has medical bills will be designated CM. If there is lost time from work, wages will be paid and medical bills and the claim will be designated CB. The database was comprised on 1,154 CM and 822 CB claims.

Job Class 9050, 8391, 8810, 4304, 9155, and 9008 account for nearly 30 percent of the claims in the database. The top 14 Job Classes, their diagnosis, and frequency are found in Table 4. A better indication of the injury may be the NCCI code. The code has the part of body, cause and mechanism of injury. The cause of an injury is identified in the NCCI code (Table 5). The most frequent injury resulted from lifting and this was followed by a fall or slip.

There were 1,183 males and 793 females in the sample with an average age of 37.58 years (SD 13.49 years). A review of the Average Weekly Wage (AWW) found that male's AWW was $469.89 (SE $12.43) and the female's AWW was $351.67 (SE $15.15) $p < 0.0001$. The wages were for all jobs and not matched by type of job but they were all from the services sector.

AWW for CB versus CM claims were statistically different p < 0.0001. CB AWW was $602.26 (ASE $14.04) and CM AWW was $292.34 (SE $11.85). AWW for CB closed and open claims were not statistically different p = 0.39. AWW for CM closed and open claims were not statistically different p = 0.61.

Discussion

The findings of our analysis demonstrate that musculoskeletal injuries are the primary cause of workers' compensation claims in the services sector. Knowing the cause of an injury can allow for changes in the environment. A fall on stairs may be secondary to poor lighting or surface problems. A fall on ice or snow may be prevented by changes in snow and ice removal. Policy changes or use of assist devices have been shown to prevent back injuries.

The difference in missing diagnostic codes (NCCI, ICD-9) is dependent upon how the information is captured. The NCCI code is a function of the case manager while the ICD-9 code is dependent upon the treating provider's billing or medical records. Improved capturing of the ICD-9 has been accomplished by allowing nurse and claim case managers the ability to enter the data into the case management system which populates the database.

Conclusion

The fragmentation and limitations of workers' compensation information systems has a direct impact on the ability to prevent illness and injury in workers' compensation.

There is a need to expand the sources of data examined in Workers' Compensation analyses. Such an expanded view will enhance our understanding of the exposures in the workplace and reduce the gulf between surveillance and information systems, and result in better interventions.

The frequency of lumbar, cervical, ankle, shoulder, and wrist conditions in the services sector identifies areas for intervention. Intervention can take the form of employee and employer education, workplace ergonomic review, surveillance programs, and enhanced safety interventions.

Recommendations

While a wealth of Workers' Compensation data exists, access is often a significant obstacle. It would be beneficial to identify information needs regarding work-related injuries and occupational diseases in the services sector. These needs could be shared with individual insurance carriers and organizations such as National Council on Compensation Insurance and Workers' Compensation Research Institute. Access to the data would be the next step in determining what workplace interventions are necessary. Even if carriers or national organizations are not willing to share their databases, it may be possible that queries be performed to answer questions posed. The type of partnership would be a first step for the various stakeholders to work together for a common cause.

References

Arday, S. L., Arday, D. R., Monroe, S., Zhang, J (2000) "HCFA's Racial and Ethnic Data: Current Accuracy and Recent Improvements," Health Care Financing Review, 21:4, pp107-116

Dennison, C., Pokras, R. (2000) "Design and Operation of the National Hospital Discharge survey: 1988

Downey, G. (2001) "Need for privacy in health sector a matter of life and death," Technology in Government, July 2001, pp 7.

Krause N, Scherzer T, Rugulier R. Physical workload, work intensification, and prevalence of pain in low wage workers : results from a participatory research project with hotel room cleaners in Las Vegas. Am J Ind Med 2005 ;48(5):326-337.

Leigh JP, Marcin JP, Miller TR. "An estimate of the U.S. government's undercount of nonfatal occupational injuries". J Occup Environ Med 2004;46:10-18.

Rosenman KD, Reilly MJ, et. al. "Cleaning products and work-related asthma", J Occup Environ Med, 2003 May: 45(5):556-563.

Workers' Compensation Data Utilization in Injury Prevention Research at the Liberty Mutual Research Institute for Safety

Theodore K. Courtney, Liberty Mutual Research Institute for Safety

Background and Introduction

The Liberty Mutual Research Institute for Safety (LMRIS) in Hopkinton, Massachusetts is a unique organization within Liberty Mutual Insurance Company. The Research Institute seeks to advance scientific, business-relevant knowledge in workplace and highway safety, and work disability. Comprised of centers focused on physical ergonomics, behavioral safety sciences, injury epidemiology, and return to work, respectively, the Institute is located in Hopkinton, Massachusetts. A unique feature of the Institute is that the results of its research projects are published in the peer-reviewed scientific press.

Research initiatives at Liberty Mutual have a storied history dating back to very nearly the founding of the company in 1912. Initial research efforts in the first half of the 20th century were consolidated into a single enterprise with the opening of the Liberty Mutual Loss Prevention Research Center in Hopkinton, Massachusetts in 1954. For the better part of the next half century, the Center's (now Institute's) work focused primarily on laboratory-based sciences and technologies including research underlying modern automotive protection technologies such as seat belts and air bags, emergency shut off switches, technologies to measure surface friction and development of manual material handling guidelines used around the world.

The purpose of this paper is to briefly introduce the Institute's historic and more current use of workers' compensation data for prevention-oriented research. In addition the paper will briefly summarize the process for the application of prevention-oriented research results of workers' compensation in the business environment.

Early Institute Research in Workers' Compensation Data

As workers' compensation claims began to be tracked and managed electronically and then converted into electronic data, the LMRIS staff recognized opportunities to harness claims data for research purposes. One of the earliest, if not the earliest, of these was field work by Snook, Campanelli and Hart (1978) in which the application and efficacy of psychophysically-derived manual material handling guidelines was assessed using field loss control assessments of low back injury cases identified from workers' compensation claims data.

In subsequent years, Snook and colleagues at LMRIS continued to work with claims data focusing on describing the costs of workers' compensation claims for musculoskeletal disorders such as those of the low back and upper extremity (e.g., Webster and Snook, 1990, 1994). These primarily descriptive studies, among the most widely cited in occupational safety research to date, paved the way for the subsequent expansion of injury epidemiology research, and later, return to work research at LMRIS.

Expansion of Institute Workers' Compensation Research

With the launch of the Liberty Mutual Research Center for Safety and Health in the early 1990's, the Institute's work in epidemiology was greatly expanded including hiring the first staff with formal credentials in epidemiology and biostatistics. From this point, workers' compensation claims data were explored on a number of fronts including describing particular subsets of injuries and their antecedents, costs and/or disability durations. As examples consider Leamon and Murphy (1995) in slips, trips and falls; Brogmus et al. (1996), Hashemi et al. (1997, 1998), Murphy and Volinn (1999) and Murphy and Courtney (2000) in musculoskeletal disorders (MSDs). Studies were also undertaken that compared workers' compensation claims data with other national injury data whether workers' compensation-based (e.g., NCCI) or not (e.g., the Bureau of Labor Statistics (BLS) annual survey)(Murphy et al., 1996). Additionally during this time, LMRIS scientists and their collaborators also began to examine the potential of workers' compensation data for looking at issues such as the recurrence of low back pain (MacDonald et al., 1997) and the use of narrative text describing claim antecedents for supplemental case identification and selection (Sorock, Ranney and Lehto, 1996)

Other efforts in the 1990's included studies using workers' compensation data to better understand health care practices in diagnosing and treating MSDs (e.g., Tacci et al., 1998, 1999; Mahmoud et al., 2000), and combining workers' compensation data with other data sources to create multi-national perspectives on occupational injury problems (e.g., in slips, trips and falls- Courtney, Sorock et al., 2001). Originally applied to injury datasets such as the National Health Interview Survey, the late 90's and early half of this decade also saw the exploration of semi-automated classification paradigms to leverage information on antecedents found in injury narratives including workers' compensation claims narratives (Wellman, et al., 2004).

At the turn of the 21st century, LMRIS launched the Center for Disability Research (CDR) and subsequently the Quantitative Analysis Unit (later to become the Center for Injury Epidemiology -CIE) which consolidated workers' compensation researchers at the Institute into roughly pre-injury/primary prevention (QAU, CIE) and post-injury/tertiary prevention/return to work-oriented groups (CDR). As the focus here is on primary prevention uses of workers' compensation data, the remainder of this paper addresses that aspect of the Institute's work. Interested readers are directed to the Institute's website (http://www.libertymutualgroup.com/omapps/ContentServer?pagename=LMGroup/Views/LMG&ft=2&fid=1138356633468&ln=en) for more information about the CDR and its extensive research programs in disability and return to work.

Workplace Safety Index
Also at this time in 2000, the Institute undertook the development of the Liberty Mutual Workplace Safety Index (LMWSI). This annual metric of the direct burden of the most severe disabling workplace injuries in the US each year was developed through a collaboration of CIE and CDR researchers. The LMWSI, which Dr. Wellman discussed earlier in the workshop, classifies the direct cost burden in terms of the putative antecedent event underlying the injury. The Index has stimulated the conversation in the research and practice communities over the past decade and served to focus attention on preventable antecedent 'mechanisms' in the workplace.

More Recent Primary Prevention Research using Workers' Compensation Data
More recently, the Center for Injury Epidemiology's primary prevention research efforts using workers' compensation claims data have included:

Descriptive studies of the construction industry, the restaurant industry, and ladder fall fractures (Courtney, Webster, and Matz, 2002, Wellman, et al., 2005; Smith et al.; 2006);

The further exploration and use of narrative text and specialized analysis/ classification frames in welding and electrical injuries (Lombardi et al., 2005, 2009);

Analytic studies, such as case control and cross-sectional studies, examining risk factors associated with fractures from same-level falls in younger and older working women (Verma, Sorock et al. 2007; Verma et al., 2008,);

Continued exploration and development of semi-automated narrative text analysis approaches using Bayesian algorithms (Lehto, Marucci-Wellman, and Corns, 2009).

How has Institute WC Injury Prevention Research been applied?

Informing Research
Within the Institute, much of the knowledge gained in these studies of workers' compensation claims data was used to guide the development of original descriptive and analytic field studies examining etiologic aspects more closely; the development of intervention studies; and the development of laboratory studies of a variety of occupational scenarios in areas such as manual materials handling; slips, trips and falls, and other sudden onset traumatic injury mechanisms. Moreover, as the results are published in the open peer-reviewed press, this workers' compensation research has contributed to the work of others and to national and international occupational health and safety initiatives and standards.

Informing Practice
Beyond publishing in the peer-reviewed press, LMRIS also reaches out to the national and international practice community through its newsletter and annual report, as well as through contributions to the professional safety and trade press.

Our top corporate loss control experts interact synergistically with research scientists at the Institute. This interaction with the research scientists allows corporate loss control experts to identify research findings for use in developing practice applications and interventions which can then be applied through our loss control staff to the specific needs of customers. Similarly, these interactions provide another avenue of data that helps inform the thinking of the research staff with respect to important workplace issues.

Beyond Research
While beyond the scope of this presentation, it would be an omission not to briefly acknowledge that there are a myriad of business uses of the workers' compensation data more typical of insurance practices in loss control and prevention. These include, but are not limited to the review of claims data experience for a particular organization or location to better understand its loss experience and alignments with potential organizational, behavioral, and engineering or environmental risks. Increasingly, such uses in loss control are being supplemented with approaches involving leading indicators (e.g., such as surveys of practices and safety climate/culture). However, these business uses of WC data are customer-centric and beyond the scope of this presentation.

References

Brogmus GE, Sorock GS, Webster BS. Recent trends in work-related cumulative trauma disorders of the upper extremities in the United States: an evaluation of possible reasons. J Occup Environ Med. 1996 Apr;38(4):401-11.

Courtney TK, Matz S, Webster BS. Disabling occupational injury in the US Construction industry, 1996. J Occup Environ Med. 2002 Dec;44(12):1161-8.

Hashemi L, Webster BS, Clancy EA, Courtney TK. Length of disability and cost of work-related musculoskeletal disorders of the upper extremity. J Occup Environ Med. 1998 Mar;40(3):261-9.

Hashemi L, Webster BS, Clancy EA, Volinn E. Length of disability and cost of workers' compensation low back pain claims. J Occup Environ Med. 1997 Oct;39(10):937-45.

Leamon TB, Murphy PL. Occupational slips and falls: more than a trivial problem. Ergonomics. 1995 Mar;38(3):487-98.

Lehto M, Marucci-Wellman H, Corns H. Bayesian methods: a useful tool for classifying injury narratives into cause groups. Inj Prev. 2009 Aug;15(4):259-65.

Lombardi DA, Pannala R, Sorock GS, Wellman H, Courtney TK, Verma S, Smith GS. Welding related occupational eye injuries: a narrative analysis. Inj Prev. 2005 Jun;11(3):174-9.

Lombardi DA, Matz S, Brennan MJ, Smith GS, Courtney TK. Related electrical injuries: a narrative analysis of workers' compensation claims. J Occup Environ Hyg. 2009 Oct;6(10):612-23.

MacDonald MJ, Sorock GS, Volinn E, Hashemi L, Clancy EA, Webster B. A descriptive study of recurrent low back pain claims. J Occup Environ Med. 1997 Jan;39(1):35-43.

Mahmud MA, Webster BS, Courtney TK, Matz S, Tacci JA, Christiani DC. Clinical management and the duration of disability for work-related low back pain. J Occup Environ Med. 2000 Dec;42(12):1178-87.

Murphy PL, Sorock GS, Courtney TK, Webster BS, Leamon TB. Injury and illness in the American workplace: a comparison of data sources. Am J Ind Med. 1996 Aug;30(2):130-41.

Smith GS, Timmons RA, Lombardi DA, Mamidi DK, Matz S, Courtney TK, Perry MJ. Work-related ladder fall fractures: identification and diagnosis validation using narrative text. Accid Anal Prev. 2006 Sep;38(5):973-80. Epub 2006 Jun 5.

Snook SH, Campanelli RA, Hart JW. A study of three preventive approaches to low back injury. J Occup Med. 1978 Jul;20(7):478-81.

Tacci JA, Webster BS, Hashemi L, Christiani DC. Clinical practices in the management of new-onset, uncomplicated, low back workers' compensation disability claims. J Occup Environ Med. 1999 May;41(5):397-404.

Tacci JA, Webster BS, Hashemi L, Christiani DC. Healthcare utilization and referral patterns in the initial management of new-onset, uncomplicated, low back workers' compensation disability claims. J Occup Environ Med. 1998 Nov;40(11):958-63.

Verma SK, Sorock GS, Pransky GS, Courtney TK, Smith GS. Occupational physical demands and same-level falls resulting in fracture in female workers: an analysis of workers' compensation claims. Inj Prev. 2007 Feb;13(1):32-6.

Verma SK, Lombardi DA, Chang WR, Courtney TK, Brennan MJ. A matched case-control study of circumstances of occupational same-level falls and risk of wrist, ankle and hip fracture in women over 45 years of age. Ergonomics. 2008 Dec;51(12):1960-72.

Webster BS, Snook SH. The cost of compensable low back pain. J Occup Med. 1990 Jan;32(1):13-5.

Webster BS, Snook SH. The cost of 1989 workers' compensation low back pain claims. Spine (Phila Pa 1976). 1994 May 15;19(10):1111-5; discussion 1116.

Webster BS, Snook SH. The cost of compensable

upper extremity cumulative trauma disorders. J Occup Med. 1994 Jul;36(7):713-7.

Wellman HM, Lehto MR, Sorock GS, Smith GS. Computerized coding of injury narrative data from the National Health Interview Survey. Accid Anal Prev. 2004 Mar;36(2):165-71.

Wellman, H.M., Filiaggi, A.J., and Courtney, T.K. Occupational injuries in US restaurants – an analysis of insurance claims data. XVII World Congress on Safety and Health at Work, 19–22 September 2005. Orlando, FL.

Using Employer Records – Pitfalls and Opportunities

Shelley Brewer, DrPH, CSP, ChemPlan, Inc., Jessica M. Tullar, PhD, The University of Texas, School of Public Health, and Benjamin C. Amick III, PhD, The University of Texas, School of Public Health, and The Institute for Work & Health

Introduction

A common problem when trying to conduct occupational research is to get "buy in" from employers. Employers are typically not open to providing the sensitive information that is often requested by researchers. In addition, researchers often add work to employers and employees by requesting they complete forms that are not a normal part of their job requirements. On the surface it seems a simple answer to make research less burdensome and more attractive to employers is to use records that the employers have available. Not only does using existing records lessen the burden on employers, it could also provide a means to lessen the burden on researchers by minimizing data collection. Unfortunately, employer records are often maintained for internal or regulatory purposes and are not easy to decipher for those outside the organization. Frequently, the published literature uses estimates of employee hours because they might not have access to more detailed information or may not understand the employer information that is provided (1). The question is "How Does Using Employee Hour Estimates Affect Reported Intervention Effectiveness?"

The methods utilized in a participatory intervention in a Dallas, Texas hospital provided the data needed to compare the use of estimated hours versus actual exposure hours. Hours were obtained from employer records. The employer involved in the study is a non-subscriber to the Texas workers' compensation system. Texas does not require all employers to maintain workers' compensation insurance. The employer managed to cut their costs for treating injured employees in half within 12 months of becoming a non-subscriber. Although seemingly contradictory, the employer's incidence rate doubled within the same period that their costs were cut in half. The exposure hours are important data to consider when evaluating how such divergent outcomes were happening at the same time. This paper will provide a quick overview of the original study followed by a description of what steps were taken to obtain exposure hours from the employer's databases. The discussion portion of the paper outlines how using estimated hours of exposure would have changed the results regarding intervention effectiveness.

Original Study

The study hypothesized that a high engagement participatory intervention would result in improvements in employee health and employee performance. The specific outcomes measured included turnover and employee injuries. Study participants were Patient Care Assistants (PCAs) and Patient Care Technicians (PCTs). An interrupted time series with a nonequivalent no-treatment control group design was developed using employer records from August 2004 to June 2007 (study period = August 2005 to May 2006 with a year pre and post intervention.) The intervention consisted of a series of facilitated group sessions focused on helping employees find meaning and purpose in their work. Employees were encouraged to identify improvements they could make both on the individual and organizational levels (www.sacredvocation.com). The intervention consisted of 12 group sessions using a Participatory Action Research (PAR) process.

Table 1. Study Sample Demographics (2)

	Participants		Non-Participants	
Demographics	N =187	%	N=561	%
Ethnicity				
American Indian/Alaskan Native	1	0.5%	4	0.7%
Asian/Pacific Islander	16	8.6%	73	13.0%
Black (Not of Hispanic Origin)	129	69.0%	244	43.5%
Hispanic/Latino	21	11.2%	79	14.1%
Not Indicated	1	0.5%	5	0.9%
White (Not of Hispanic Origin)	19	10.2%	156	27.8%
Gender				
Male	17	9.1%	117	20.9%
Female	170	90.9%	444	47.6%
Age Range				
20-29	38	20.3%	276	49.2%
30-39	43	23.0%	165	29.4%
40-49	55	29.4%	78	13.9%
50-59	38	20.3%	31	5.5%
60-69	13	7.0%	10	1.8%
70-79	0	0.0%	1	0.2%
Job Tenure				
less than 3 years	40	21.4%	362	64.5%
3 to 6 years	55	29.4%	124	22.1%
7 to 10 years	25	13.4%	20	3.6%
11 to 20 years	41	21.9%	32	5.7%
21 to 30 years	18	9.6%	18	3.2%
31 or more years	8	4.3%	5	0.9%

Table 2. Injury Cause (2)

Injury Cause Code	Pre-intervention	Post-intervention	Total	%
Bodily Exposure	12	13	25	23%
Caught In, Under, Between	0	1	1	1%
Combative Patient	0	2	2	2%
Cut-Puncture-Scrape	2	10	12	11%
Miscellaneous	1	1	2	2%
Slip – Trip – Fall	2	7	9	8%
Strain	25	24	49	46%
Struck-by	5	2	7	7%
Total	47	60	107	

Table 3. Employer Raw Data

Entry	Employee ID	Time Period	STI	PTO	Reg	OT	**Total Hours:**
6984	62	24		24	56.1	0.75	**80.85**
6985	62	25		34.93			**34.93**
6986	62	26	160	-34.93			**125.07**
6987	62	27	80				**80**
6988	62	28	80				**80**
		Total:	320	24	56.1	0.75	**400.85**

The employer records obtained to determine the outcome measures were Occupational Safety and Health Administration (OSHA) logs and a Human Resources (time keeping) database. The Human Resource database included hours worked, age, gender, ethnicity, unit worked and employment start date. A second database provided by risk management included injury dates, type of injury and cause of injury. A third database was developed by the principal investigator to track intervention participation. The three databases were linked using a de-identified number to track employees.

The final sample used in the analysis included 748 individuals with a total of 20,674 hours worked entries (from Human Resource database). The 748 individuals included 187 participants and 561 non-participants (Table 1). One individual could have more than one entry per pay period

Figure 1. Codes for Hours Worked on Floor

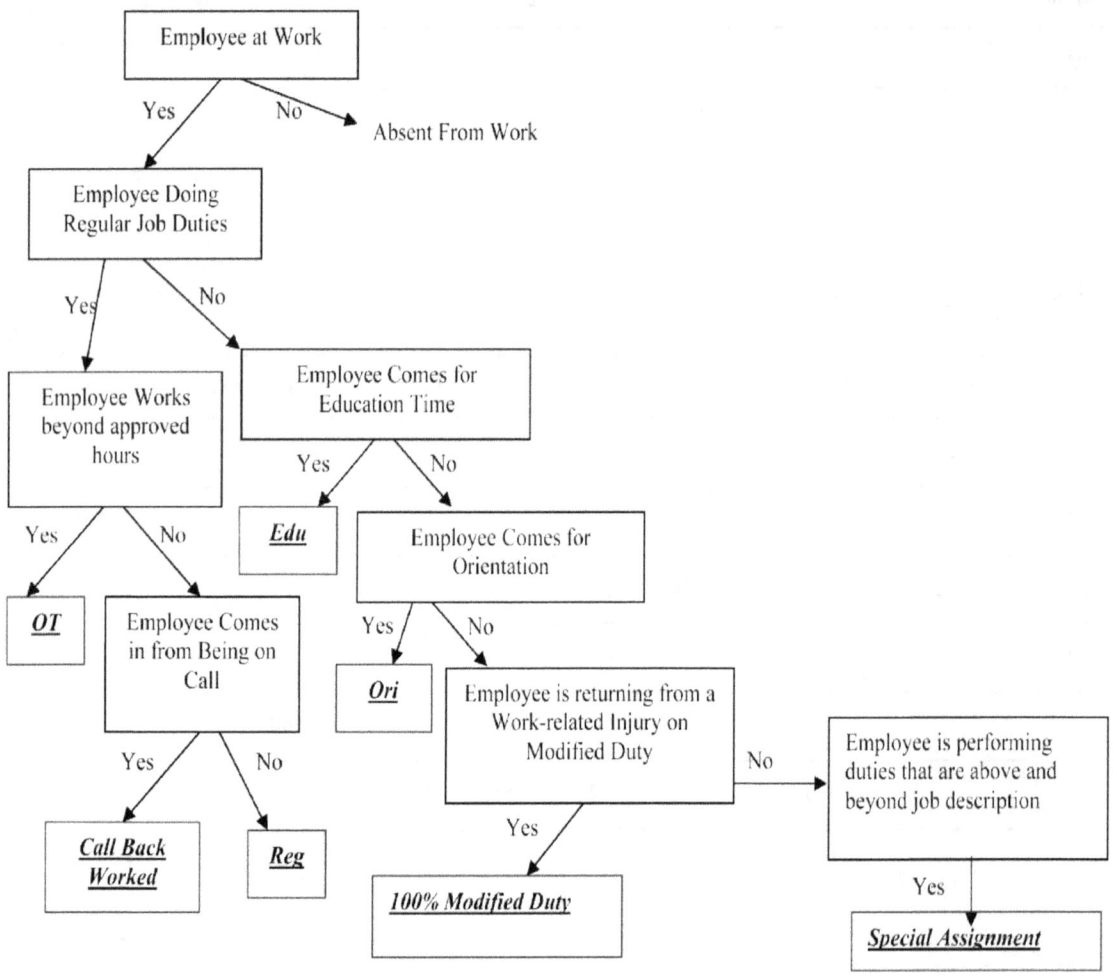

Table 4: Pay Code Categories

Pay Code Groups		
Hours on Floor	**Hours Absent**	**Benefit Hours**
Regular Earnings	Paid Time Off Usage	Adoption Assistance @ 60% Pay
Straight Overtime @ 1_0	Sick Pay	Award Pay
Education In Training	100% Paid Short Term Illness	Bonus
Modified Duty 100%	60% Paid Short Term Illness	Clinical Transformation
Modified Duty 90%	Baylor Safe Choice 90%	Funeral Pay
On Call Pay	Intermittent FMLA – Sick	Incentive Date Verification
Call Back Worked	Intermittent FMLA – Unpaid	Inclement Weather
Retro Base w/401K	Intermittent FMLA-PTO	Jury Duty Pay
Orientation Pay	Paid Partial Disability w/FMLA	Shift Three Night Differential
	Short Term Disability	

due to the method the hospital recorded hours worked. One hundred and seven injuries were reported from the 748 employees (2).

The most common injuries reported were strains and sprains followed by bodily exposures (Table 2). The mean injury cost for the entire study period was $2,319 with a range of $0 to $34,168. Post-intervention injury costs (mean = $1,397) were less than half of pre-intervention costs (mean=$3,496).

Methods Used to Obtain Exposure Hours:
The employer provided the hours worked for each employee from their Human Resource database. The entries were reviewed and common errors such as missing data and transposing of numbers were observed. The common errors were easily cleaned as the mistakes were not unique to the hospital database. The observation of items such as negative work hours, multiple entries and unknown pay codes were more troubling. In addition, the Risk Management database that tracked injuries did not identify employees the same way the Human Resource database identified employees. The exposure hours needed to be identified along with linking these hours to the correct employee in the Risk Management database. The first step to "cleaning" the data was to learn the difference between administrative codes used to ensure employees received the correct pay and codes that indicated "hours on the floor" (exposure hours). Items such as negative hours and duplicate entries had to be identified and corrected. Table 3 provides an example of typical raw data entries.

Table 5: Incident Rates

	Non-Participants	Participants	All Employees
Pre-Intervention			
IR using Exposure Hours	9.49	16.30	12.78
IR using Raw Data	10.72	21.06	15.70
IR using Estimated Hours*			8
Post-Intervention			
IR using Exposure Hours	19.22	16.14	17.81
IR using Raw Data	34.08	18.34	26.87
IR using Estimated Hours*			13

*~number of employees X 40 hours/week X 50 weeks/year

Two decision trees were developed as the researchers worked with the employer to learn which pay codes (such as PTO (paid time off), Reg (regular), OT (overtime), etc.) should be used to calculate exposure hours. Figure 1 identifies the codes (in bold print and underlined in Figure1) used to indicate hours worked on the floor. The decision tree for codes indicating hours absent from work is not included due to size.

The examination of the pay codes resulted in the employee hours being separated into three major categories (Table 4).

After the exposure hours were determined, the employees were linked to the injury and intervention databases.

Results – Comparing Exposure Hours to Estimate Hours

The original study utilized logistic regression and multi-level modeling to determine the effectiveness of the intervention. The results in this paper present the OSHA Incidence Rate calculation commonly used in the workplace. We were interested to see if the time spent cleaning the data to achieve true exposure hours was worth the effort. Will the reported incidence rate be different when using exposure hours instead of an estimate or the raw data?

The OSHA Recordable Incident Rate (IR) is calculated by multiplying the number of recordable cases by 200,000 and then dividing that number by the total labor hours (IR= Number of OSHA Recordable Cases X 200,000/ Total Number of Labor Hours). The standard base rate is based on a rate of 200,000 labor hours. 200,000 is equal to 100 employees who work 40 hours per week 50 weeks per year (3). The use of a standardized base rate allows companies to calculate their rate(s) and get a percentage per 100 employees (Table 5). The calculation is more meaningful for larger companies as smaller companies have fewer labor hours which results in higher rates and greater fluctuations from year to year. The OSHA IRs for both pre and post intervention are listed below:

The calculation using the "estimated" number of employees is closer to the rate resulting from the true exposure hours in this worksite. However, the IR using the estimated hours could have been almost twice the true exposure hour IR if different assumptions were made. The estimated hours in this example was made using 600 employees (a FTE estimate) instead of the 748 because the records indicate the whole study population was not counted for 50 weeks. The estimated IRs could have been greatly reduced or increased depending on the estimate of employees used to create the estimated man hours.

Discussion

Researchers use a wide range of outcome measures when reporting intervention effectiveness (4). Often, estimates of full time employees are used in calculations to determine exposure hours. The information presented in this paper demonstrates that estimates of FTE and "unclean" data may result in both overestimation and underestimation of intervention effectiveness.

This project also demonstrates that employer records can be used to ascertain individual level data. The process used to determine exposure hours vs. hours paid was a painstaking process that involved the employer, researchers and individuals experienced in workplace policies and practices. The process was worth the effort as we were able to determine that the difference in exposure hours vs. the raw data/estimated FTE was large enough to change the reported effects if the IR was the only outcome measure reported.

Intervention researchers need to understand the language used in workplaces and OSHA rates are part of this language. Scientists can improve on the accuracy of the OSHA rates by using exposure hours instead of estimates. Researchers should also spend the time to understand and work through employer data when it is available. Often researchers will say this information is unreliable and has validity concerns because it is self reported. Somehow employer records are considered unreliable but self reports of pain or discomfort on a survey are considered reliable because science has validated the instrument (5). While this conclusion makes sense to researchers, employers may not understand why their records are not considered usable. Scientists need to consider how to use employer records validly rather than simply discounting them.

Another issue this project exposes is insurance status can drive what employers find important. The hospital in this project was a "non-subscriber". The employer's focus was on reducing the cost of treating injuries, not on reducing the incident rate. Researchers need to understand the drivers. The fact the employer's incident rate went up is not surprising if you know the employer instituted a new policy mandating the reporting of all injuries/incidents. We do not know whether employees were at greater or less risk of injury given the change in the reporting system but it is important to mention when reporting effects.

The issues raised in this paper demonstrate that employers and scientists need to work together to move occupational research forward. Both audiences need to realize that progress in employee safety is most likely to occur when we educate each other on what we know.

References

1. Amick, B., Tullar, J., Brewer, S., Mahood, Q., Irvin, E., Pompeii, L., et al. (2006). Interventions in health-care settings to protect musculoskeletal health: A systematic review. Toronto: Institute for Work & Health.

2. Tullar, JM. (2008). An epidemiologic evaluation of a worksite intervention. Ph. D. Dissertation. Houston: University of Texas School of Public Health.

3. Injury Calculations. (2007). www.osha.gov.

4. Cullen, et al. (2006) Use of Medical Insurance Claims Data for Occupational Health Research. J Occup Environ Med; 48: 1054-1061.

5. Brewer, S. (2007). Workplace Injury/illness Prevention and Loss control programs: A Series of Systematic reviews. DrPH Dissertation. Houston: University of Texas School of Public Health.

Multi-Agency Data Matching to Detect Suspected Uninsured Employers: Research Impacts Policy

Christine Baker and Chris Bailey, California Department of Industrial Relations

The experience in California of using administrative data from different agencies demonstrates that important policy questions can be answered to improve the health and safety and workers' compensation systems. The California experience also demonstrates that while interagency cooperation and data sharing are possible, it is not guaranteed without clearly formulated legislation. In the case of identifying illegally uninsured employers, it is suggested that many more non-compliant employers could be identified, investigated and fined by the agency if resources and priorities would allow.

Background

All employers in California except the State (which is permissibly uninsured) are required to provide workers' compensation coverage for their employees through the purchase of workers' compensation insurance or by being certified by the State as permissibly self-insured. In order to treat injured workers whose employers may have illegally cheated by not purchasing workers' compensation insurance, the Uninsured Employers Benefits Trust Fund (UEBTF) is available to fill the gap until payment responsibilities are properly adjudicated and reimbursed. However, access to this fund is complicated and can be untimely, and utilization of the normal route of workers' compensation insurance is how the system is designed to function by law.

Non-compliance by illegally uninsured employers and other types of insurance fraud are addressed in various sections of the Labor and Insurance codes. Investigators and inspectors use penalties and prosecutions to enforce these laws. However, in the previous decade, the workers' compensation community recognized that enforcement resources had not eliminated (significantly reduced?) fraudulent activities in the system. In February 1997, the California Commission on Health and Safety and Workers' Compensation (CHSWC or "the Commission") conducted a public fact-finding hearing in Los Angeles on workers' compensation anti-fraud activities and, among other findings, determined that some employers were not complying with the requirement to secure workers' compensation coverage for their workers.

In April 1997, CHSWC released a draft report on the Los Angeles public fact-finding hearing and in September 1997, released the final report with recommendations, including input from the Department of Insurance (CDI). The final report makes reference to the data matching project and an Issue paper with recommendations.[1]

The Commission then voted to engage in pilot data matching projects in 1998 and to create a CHSWC Uninsured Employer Roundtable to consider and suggest Legislative changes. In December 1998, the Commission approved the report findings and recommendations from the data matching pilot.[2]

Data Matching Pilots

A series of pilot studies were conducted in 1998 to attempt to identify illegally uninsured employers and bring them into compliance. The stated goals of this project were multiple, including a proof-of-concept exercise, an extrapolation estimate, and applied approaches to cost savings, injured worker protection and enforcement. Specifically, the goals were to:

---
[1] http://www.dir.ca.gov/chswc/Fraud/Fraudreport.html
[2] http://www.dir.ca.gov/chswc/uefcover.html

- Protect workers from lack of workers' compensation coverage;
- Identify illegally uninsured employers and bring them into compliance;
- Reduce the cost to the state's UEBTF and General Fund;
- Reduce the need of workers who are injured while working for illegally uninsured employers from using other social and benefit systems;
- Level the economic playing field for insured employers;
- Protect the State from increased liability faced by UEBTF;
- Determine the effectiveness and cost-benefit of a matching records program to identify illegally uninsured employers and bring them into compliance.

The pilot project used data matching techniques to compare coverage records from the California Workers' Compensation Insurance Rating Bureau (WCIRB), the only entity in California which has a complete list of employers with coverage, with data from the Employment Development Department (EDD) which captures reported unemployment insurance and payroll.

Each pilot project targeted a specific group of employers. (See http://www.dir.ca.gov/chswc/uefintro.html and Appendix A.) One pilot tested the incidence of established employers who once had workers' compensation coverage and were established enough to be experience-rated (X-mod), but who then allowed themselves to not renew their policy or let their coverage lapse. Another pilot tested the incidence of certain industries suspected of high rates of non-compliance such as auto repair and restaurants, as well as the incidence of non-compliance in the general population of employers. The last pilot tested the incidence of non-compliance for new employers. All of the test pilots drew samples from the EDD database. All of the pilots involved follow-up by the Division of Labor Standards Enforcement (DLSE) either in the form of a notice or on-site inspection, or both. In the pilot targeting industries, WCIRB initiated the first letter and DLSE followed up with a notice from non-respondents.

The results of the pilots proved to CHSWC that the issue of illegally uninsured employers was indeed a serious problem and a widespread one based on the different targeted test groups. The experience-rated pilot results nullified the hypothesis – not one experience-rated employer sampled was found out of compliance[3], suggesting that employers who succeed in becoming established play along with at least the most basic rules of tax and insurance law, that is, compliance, but not necessarily accurate reporting. Overall findings concluded that approximately 9% of employers at large and nearly 20% of employers in select industries were illegally uninsured. Further, the follow-up activities were effective at achieving employer compliance actions.

---

[3] All employers were either insured, out-of-business, or had no employees – meaning they did not require coverage.
[4] Since the year the EDD sample was drawn, the California Labor and Workforce Development Agency (LWDA) was created, uniting the Department of Industrial Relations (DIR) and EDD under the same agency Secretary.
[5] For example,
EEEC sweeps in 2005 http://www.dir.ca.gov/DIRNews/2005/IR2005-40.html, http://www.dir.ca.gov/DIRNews/2005/IR2005-32.html,
EEEC sweeps in 2006 http://www.dir.ca.gov/DIRNews/2006/IR2006-12.html, http://www.dir.ca.gov/DIRNews/2006/IR2006-22.html,
EEEC sweeps in 2007 http://www.dir.ca.gov/DIRNews/2007/IR2007-13.html, http://www.dir.ca.gov/DIRNews/2007/IR2007-12.html,
http://www.dir.ca.gov/DIRNews/2007/IR2007-24.html,
http://www.dir.ca.gov/DIRNews/2007/IR2007-51.html;
other joint sweeps in 2004, before the EEEC name at http://www.dir.ca.gov/dirnews/2004/IR2004-15.html, http://www.dir.ca.gov/dirnews/2004/IR2004-6.html, http://www.dir.ca.gov/dirnews/2004/IR2004-16.html.

With the pilots completed, in December 1998, CHSWC released a report which detailed the methods, tabulated the data results, and provided recommendations based on the pilot. Recommendations included how to improve the data matching process, to consider making the project on-going, and to institute penalties on uninsured employers for failing to respond to notices, among other recommendations. The report acknowledged EDD among those of assistance in the project. Clearly, the pilots could not have been conducted without the cooperation of WCIRB and EDD, which was at the time situated in a separate agency.[4]

Pilot Data Matching Results Lead to New Legislation

The results of these pilot projects provided an impetus in 2002 to create Labor Code 90.3. Labor Code 90.3 (Assembly Bill 749, Charles Calderon) created a program "for targeting employers in industries with the highest incidence of unlawfully uninsured employers" and specified multi-agency/multi-organization data sources to be used. The law also required annual reporting to the Legislature on the effectiveness of the program. Incidentally, the large bill which created Labor Code 90.3 also required the Bureau of State Audits to evaluate all levels of fraud fighting efforts by DIR and CDI.

Several attempts were made to match records, but very little follow-through occurred. Changing management in DLSE resulted in a lack of resources being dedicated to the program. Further, due to a lack of enabling funding authority, the program was not implemented systematically, and the previously mentioned pilot projects served as the only quantitative evidence of the effectiveness of multi-agency/multi-source data matching methodologies to detect uninsured employers. Meanwhile, throughout the mid-2000s, CHSWC attempted to update the data matching project with cooperation from EDD. Since the time of the first pilot project, EDD had become united with DIR under the same agency. Memoranda of Understanding were drawn up in the period leading up to 2006, but ultimately, cooperation in this data sharing exercise failed this time around.

In April of 2007, CHSWC released a UEBTF (mentioned above in the Background section) report which, among other issues, detailed the costs and recoveries of the fund. The report demonstrated that despite the complexity and difficulty of accessing the fund, the fund was nevertheless relying primarily on assessments rather than penalties and recoveries for their budgeting. This fact led some stakeholders and policy advocates to the emphasize improved efforts and results from DLSE whose responsibility is to identify, investigate and fine illegally uninsured employers.

During this same period, DLSE and others were working with the Economic and Employment Enforcement Coalition (EEEC), a multi-agency enforcement program consisting of investigators from DLSE, Division of Occupational Safety and Health (DOSH), EDD, the Contractors State License Board (CSLB) and US Department of Labor (DOL), which was created to "combine the enforcement efforts of the agencies and put as many investigators into the field as possible." Press releases of the results of EEEC detailed the targeted "sweeps" of businesses in various industries and in various geographical areas.[5]

At about the same time in 2007, State Senator Mark Ridley-Thomas sponsored Senate Bill (SB) 869 which attempted to address the short-comings of a targeted enforcement program already in statute, Labor Code 90.3. The bill had the support of the small business group, Small Business California, as well as the California Labor Federation (AFL-CIO).

Given the history of the data matching pilot informing legislators about the efficacy of the methodology in identifying illegally uninsured employers, the resulting creation of a program in Labor Code 90.3, and then the lack of implementation of the program, authors of SB 869 as amended, crafted the legislation to provide more exact guidance on implementation procedures and department cooperation in data matching and reporting requirements, as well as clarity about authorization for funding the program.

In the fall of 2007, Governor Schwarzenegger signed SB 869 which amended Labor Code 90.3 to further specify and require a program that "systematically identifies unlawfully uninsured employers" and allowed for prioritized targeting methods, as well as other methods such as random sampling. Labor Code 90.3 provided the needed enabling funding language and refined the type of annual reporting to the Legislature (and to the public via the Internet) on the effectiveness of the program. The reporting requirements help guide the type of program that needs to be conducted. For example, the terms "matching records" and "matched to records" are used in order to direct how to report statistics, methodologies and measureable results; reported statistics should "permit analysis and estimation of the percentage of unlawfully uninsured employers that do not report to the EDD."

The statute states that the report need "not be limited to" the specified required numerical results listed in Labor Code 90.3(d)(1)-(8). Therefore, the program is given the latitude to innovate, expand and pilot new methods as needed. Despite this, as discussed below, the first report submitted could be improved by reporting on "the number of employers identified as unlawfully uninsured from records of the UEBTF or from records of the DWC (Division of Workers' Compensation)...," by making the percentage estimation of unlawfully uninsured more obvious with the inclusion of a percentage figure in the report findings, and by describing in greater detail the methodology, work process and FTEs/resources used.

Results from Labor Code 90.3 Data Matching Program and Report

In June 2009, the first report required under the amended Labor Code 90.3 was released by DIR. The report is available at: http://www.dir.ca.gov/dlse/UUEEP-2008.pdf and is reprinted in the Appendix B.

According to the report, in order to implement the systematic unlawfully uninsured employer enforcement program, a new process for multi-agency data matching was established. Through the new process, each quarter, DLSE receives from EDD a randomly selected list of 500 employers from EDD's database of reporting employers. All 500 employers for the represented quarter are reviewed by WCIRB for evidence of insurance coverage.

The reported results of the program yielded 123 citations issued per Labor Code section 3722(a) for not being insured and 33 citations issued per Labor Code section 3722(b) for being found to not be insured in the past, totaling 156 citations out of a total of 1,500 sampled employers. Of the $484,489 in workers' compensation penalties assessed, $151,783 in workers' compensation penalties were collected and $76,000 in citations were administratively dismissed (Appendix B, Table 1). While no percentage calculation is reported in the summary findings, it is estimated that 10%-12% of the sampled employers were found to be uninsured.

The report concludes that the efforts have yielded positive results in DLSE's continued work in combating the underground economy and that DLSE will continue to refine the efficiency and effectiveness of this program for the benefit of both employees and employers.

The Future of Labor Code 90.3 and the Labor Code 90.3(d) Report – Lessons Learned for Future Reporting

The data matching program is now required and funded. DLSE proved able to obtain sample data from EDD. Although the program is running, documentation regarding the findings and methods could be improved. In addition to reporting on the percentage of uninsured employers in future yearly reports, it would be useful to also characterize in the findings the industry types and sizes of employers found to be out of compliance or unresponsive.

Funded by user fee assessments, such a matching records program saves enforcement resources, targets the employers who are not in compliance and is the most effective and efficient method for identifying employers who are not insured. As the program establishes itself, it is expected that the procedures will become more routine and the process more streamlined, thereby facilitating the preparation of future yearly reports.

DLSE will always need to balance its investment in resources in fulfilling the requirements of Labor Code 90.3 with other routine, complaint-driven investigations as well as with collaborative activities with the EEEC which is focused on the broader "underground economy." The future of the program is undeniably more secure with the 2007 amendments to Labor Code 90.3, and it will be more valuable when the program provides more documentation for review by the legislature, the Executive branch and the public-at-large. In addition, the support of labor and management is critical and provides a useful balance when designing a new program.

Conclusion

This analysis of a past pilot project in California provides insight into how collaborative data sharing among government units and sound research methods can yield important results to inform the political process. In this example of data matching to detect and bring into compliance uninsured employers, research results, stakeholder discussions and proposed legislative recommendations eventually led to a statutory program to continue a matching project as a means to an improved enforcement and compliance effort.

Competing demands and resources revealed legislation without an enabling funding source and the need for further amendments to correct those deficiencies in statutory language. The program became operational in 2008, with the first required report submitted in June 2009. Observers report that program findings of the employer population which is illegally uninsured fall within a range of 10-12%. The report itself discloses that from a sample of 1,500 employers, $151,783 in penalties were collected. With some refinement in the program, there is every expectation that more employers will be brought into compliance, collections from penalties will increase, and the public will better understand the effectiveness of the program.

We would like to acknowledge the leadership of DIR Director John Duncan, the research methods and analysis by Frank Neuhauser at UC Berkeley, and the contributions by CHSWC staff Irina Nemirovsky and Chris Bailey in tracking the progress of data matching research leading to policy impacts. We recognize the California Federation of Labor (AFL-CIO) in its efforts to bring employers into compliance, and Scott Hauge, Small Business California for advocating for a "level playing field."

(Appendices follow.)

APPENDIX A.

1998 Pilot Studies

A series of pilot studies were conducted in 1998 to identify illegally uninsured employers and bring them into compliance. Each pilot project targeted a specific group of employers. (See http://www.dir.ca.gov/chswc/uefintro.html)

Pilot 1

The first pilot followed-up on a sample of 350 experience-rated employers for who policy coverage could not be identified at the time of recalculation of experience modification. Policy information was requested from the employer. Each employer that failed to respond (120) with confirmed policy information was matched to EDD records of reported wages. In April 1998, fifty-eight of the 120 were identified as apparently uninsured employers with reported wages. These employers were referred to DLSE for follow-up notification and investigation. This pilot concentrated on a sample of employers whose X-mod calculations were performed during April 1997.

Pilot 2

The second pilot targeted several industries that are responsible for a disproportionate demand upon the state General Fund through claims against the UEF. These industries are also suspected of high rates of noncompliance with the requirement that employers secure the payment of compensation.

Using specific industries (Auto/Truck Repair, Restaurants/Bars), the Employment Development Department (EDD) created random samples of 250 employers in each targeted industry based on primary SIC code. In addition, a random sample of 250 firms was drawn from all other employers.

The Workers' Compensation Insurance Rating Bureau (WCIRB) then matched these employers to policy information. During April and May of 1998, the WCIRB notified each employer with apparent lapses in coverage or for whom coverage could not be determined, requesting an explanation for current or historical lapses in coverage.

The Division of Labor Standards Enforcement (DLSE) followed-up with a mail request for policy information (Form 601). For those employers (221) who failed to respond or failed to adequately demonstrate insurance coverage, DLSE attempted on-site inspection.

Pilot 3

Pilot 3 tested methods of improving new employers' knowledge of the need for compensation coverage and identification of new employers who willfully avoid compliance. Information received from other states indicates that 40% of new employers fail to secure payment of compensation. Efforts by other states have also been very successful at bringing new employers into compliance.

EDD drew a sample of 500 new employers that reported wages for the first time in the second quarter of 1997. Each of these employers was matched to policy data by the WCIRB. All apparently uninsured employers were sent a copy of the notification that will appear in EDD's "California Employer's Guide" and a request to provide policy information or reason that insurance was not required.

A second notice was sent to all employers who did not respond. This notification reiterated the requirements for insurance, reminded the employer that lack of coverage is a crime, and informed the employer that if policy information is not returned, the employer's identity would be turned over to DLSE for follow-up which could result in penalties of up to $10,000 per employee.

If no response was received to the second notice, DLSE followed-up in the same manner as for the targeted employers in Pilot 2. DLSE followed-up on 208 employers in Pilot 3. For each employer in the pilots the following information was recorded:

- Quarterly payroll and employment
- If they had insurance
- If they responded to first notification
- If they responded to second notification
- If they were exempt, out of business, or had no payroll
- If they obtained insurance after 1st notification, but prior to second
- If they obtained insurance after 2nd notification, but prior to referral to DLSE
- The premium for most recent period available
- Their X-mod, most recent period available
- If exempt, reason

If not covered:
- penalties assessed
- penalties collected

This information would permit estimation of the percent of, number of, employment at, and premium avoided by:

- Ex-mod rated employers out of compliance
- Target industry employers out of compliance
- New employers out of compliance
- All employers out of compliance

The pilots also tested the ability for notification to bring employers into compliance without the need for penalties and investigations. Many employers, especially new employers may be unaware of the need for compensation insurance. Other employers who are temporarily out of compliance may be encouraged to obtain and maintain insurance if they are aware that compliance is being enforced.

APPENDIX B.
2008 Annual Report of the Unlawfully Uninsured Employer Enforcement Program
Labor Code Section 90.3(d)

The Division of Labor Standards Enforcement (DLSE), through its Bureau of Field Enforcement Unit (Bureau), is charged with enforcing laws requiring employers in the state of California to secure the coverage of workers' compensation insurance for any and all employees. In furtherance of this, Labor Code §90.3(b) requires the Labor Commissioner to maintain a program designed to systematically identify potentially unlawfully uninsured employers through data matching efforts involving the Uninsured Employers' Benefits Trust Fund (UEBTF), the Employment Development Department (EDD), and the Workers' Compensation Insurance Rating Bureau (WCIRB). Labor Code section 90.3(d) requires the Labor Commissioner to report annually to the Legislature concerning the effectiveness of this program.

Background

Assembly Bill (AB) 749 (Chapter 6, Statutes of 2002) added Labor Code section 90.3, mandating that the Labor Commissioner establish and maintain a targeted unlawfully uninsured employer enforcement program. Effective January 1, 2008, SB 869 (Chapter 662, Statutes of 2007) created a funding mechanism for implementation of the unlawfully uninsured employer enforcement program, and additionally modified the data reporting relationships between the partner agencies in support of the program. In implementing SB 869, the various agencies immediately worked to establish a coordinated system of data collection which includes development of detailed processes for the identification of employers, transmission and sharing of information, verification of information including notification to employers and cross-referencing of data, and inspection and enforcement against uninsured employers.

Program Results

In order to implement the systematic unlawfully uninsured employer enforcement program, a novel data collection system was established. Through the new process, each quarter DLSE receives from EDD a randomly-selected list of 500 employers from EDD's database of reporting employers. All 500 employers for the represented quarter are screened through the WCIRB for evidence of insurance coverage.

DLSE initially received data in May 2008, reflecting employer information from records for the fourth quarter of 2007 (October – December 2007). Because the implementation of SB 869 is still in its early stages and there is a delay in the data reporting, this report reflects information available as of January 27, 2009. However, some investigations based on data obtained during the prior three quarters are ongoing. [The following tables] Table 1 summarize results of this program.

DLSE provided a list to the WCIRB of 71 employers who, during the course of DLSE's investigation for the first two quarters of reported data, were able to provide proof of insurance coverage for the period of time in which the WCIRB indicated there was no coverage. [The following] Table 2 summarizes WCIRB's responses for these employers:

[The following table] Table 3 summarizes the nature of responses received from employers in accounting for a lack of workers' compensation coverage:

As shown, DLSE has undertaken significant efforts to implement the systematic unlawfully uninsured employer enforcement program, and those efforts have yielded positive results in DLSE's continued work in combating the underground economy. DLSE will continue to refine the efficiency and effectiveness of this program for the benefit of both employees and employers.

Respectfully Submitted,
Angela Bradstreet, State Labor Commissioner

Table 1

Employers identified from records of EDD that were screened for matching records of insurance coverage or self-insurance.	1,500
Employers identified from records of EDD that were matched to the records of insurance coverage or self-insurance.	949
Employers identified from records of EDD that were notified that there was no record of their insurance coverage.	551
The number of employers responding to contact for verification. (See Table 2 for nature of responses.)	279
Inquiries returned by the post office as undeliverable.	33
Employers responding who verified they had workers' compensation insurance.	71
WCIRB did not have sufficient database information on the business to respond.	27
Employers acknowledging lack of workers' compensation insurance.	138
Employers investigated.	551
Number of citations issued per Labor Code section 3722(a). [1]	123
Number of citations issued per Labor Code section 3722(b). [2]	33
Amount of workers' compensation penalties assessed.	$484,489
Amount of workers' compensation penalties collected.	$151,783
*Amount of citations administratively dismissed. [3]	<$76,000>

[1] Labor Code section 3722(a) provides: At the time the stop order is issued and served pursuant to section 3710.1, the director shall also issue and serve a penalty assessment order requiring the uninsured employer to pay to the director, for deposit in the State Treasury to the credit of the Uninsured Employers' Benefits Trust Fund, the sum of one thousand dollars ($1,000) per employee employed at the time the order issued and served, as an additional penalty for being uninsured at that time.

[2] Labor Code section 3722(b) provides: At any time that the director determines that an employer has been uninsured for a period in excess of one week during the calendar year preceding the determination, the director may issue and serve a penalty assessment order requiring the uninsured employer to pay to the director, for deposit in the State Treasury to the credit of the Uninsured Employers' Benefits Trust Fund, the greater of (1) twice the amount the employer would have paid in workers' compensation premiums during the period the employer was uninsured, determined according to subdivision (c), or (2) the sum of one thousand dollars ($1,000) per employee employed during the period the employer was uninsured. A penalty assessment issued and served by the director pursuant to this subdivision shall be in lieu of, and not in addition to, any other penalty issued and served by the director pursuant to subdivision (a).

[3] A citation may be administratively dismissed if the employer had no proof of workers' compensation insurance at the time of the inspection/citation but proof of insurance at the time of inspection was submitted later. In such a case, the dollar value of the citation as issued is counted as "penalties assessed" but the dollar value of the assessment is uncollectable.

Table 2

WCIRB found coverage under another name and/or address provided by DLSE in the follow-up lists.	29
Coverage was found by DLSE, but the policy either incepted or was not received by the WCIRB until after the date of submission of the original quarterly lists.	10
Employer was not required to have workers' compensation insurance (i.e., employer was self-insured, had no employees subject to workers' compensation requirements, or was otherwise legally not insured).	7
DLSE reported finding coverage, but did not provide sufficient coverage information in the follow-up lists and the WCIRB could not confirm coverage.	25

Table 3

Company out of business	52
No employees	49
Corporate officers only	30
Self-Insured	4

State-based Occupational Injury and Disease Surveillance

Robert Harrison, MD, MPH, Jennifer Flattery, MPH, California Department of Public Health

Role of State-Based Occupational Surveillance Systems

State-based health departments, which have the legal authority to require disease reporting and collect other health data, play a crucial role in public health surveillance (Davis 2005). According to the National Institute for Occupational Safety and Health (NIOSH) Strategic Surveillance Plan, the long-range vision of a comprehensive nationwide occupational health surveillance program is for all states to have the core capacity to conduct surveillance of occupational injuries, diseases, and hazards that will contribute to State and local prevention efforts, as well as to national data concerning magnitude, trend, and distribution. In addition, states should also have the capacity to conduct focused in-depth surveillance, follow-up investigations, and intervention for selected, targeted conditions (diseases, injuries, or hazards) (NIOSH 2010). State health agencies have several important roles in the surveillance of occupational diseases, injuries, and hazards, including providing critically needed data on occupational disease and injury; actively linking surveillance findings with intervention efforts at the State and local level; and integrating occupational health into mainstream public health practice. The Council of State and Territorial Epidemiologists (CSTE) has recently updated guidelines for a minimum level of capacity in occupational illness and injury prevention within the public health infrastructure at the state level (Stanbury 2008).

Over the past 25 years, the concept of the sentinel health event has become an integral component of stated-based surveillance. Originally described by Rutstein in 1983, a sentinel health event "is a preventable disease, disability, or untimely death whose occurrence serves as a warning signal that the quality of preventive and/or therapeutic medical care may need to be improved" (Rutstein 1983, Rutstein 1984). Rutstein described the Sentinel Health Event (Occupational) (SHE(O)) as "a disease, disability, or untimely death which is occupationally related and whose occurrence may: 1) provide the impetus for epidemiologic or industrial hygiene studies; or 2) serve as a warning signal that materials substitution, engineering control, personal protection, or medical care may be required." The original SHE(O) list encompassed 50 disease conditions that are linked to the workplace for which objective documentation of an associated agent, industry, and occupation exists in the scientific literature. The SHE(O) concept was implemented by NIOSH in 1987 as a cooperative state-federal effort (referred to as Sentinel Event Notification System for Occupational Risks or SENSOR) to use targeted sources of sentinel providers to recognize and report selected occupational disorders to a state surveillance system (Baker 1988, Baker 1989, Matte 1989).

The evolution of the original SENSOR program has moved beyond that of a single surveillance system relying solely on the sentinel provider, and currently incorporates varied sources of case ascertainment as well as active interventions. The concept of using multiple data sources to ascertain cases and implement interventions to prevent work-related injury and disease has been successfully implemented in many states for pesticide-related illness, carpal tunnel syndrome, amputations, dermatitis, burns, youth injury, carbon monoxide poisoning, tuberculosis, work-related asthma, silicosis, and severe traumatic injuries.

The Bureau of Labor Statistics (BLS) Survey of Occupational Injuries and Illnesses (SOII) system has been the underpinning of epidemiologic surveillance of workplace injuries and illnesses in the U.S. since 1972. Many attributes of this system make it ideal for tracking workplace injuries and illnesses, including the comprehensive nature of reporting, sampling characteristics that can generate state-based data, and annual rates that can be used to evaluate the impact of interventions over time. However, recent studies have suggested that the SOII can be supplemented by additional ascertainment of cases using state-based data sources. There is currently a focus on the need to improve surveillance of non-fatal injuries and both fatal and non-fatal work-related illnesses in the U.S. A recent study of injury and illness reporting in Michigan found that the BLS SOII missed more than two-thirds of job-related injuries and illnesses (Rosenman 2006), while another study estimated that the SOII missed between 33% and 69% of all injuries (Leigh 2004). Additionally, major changes in Occupational Safety and Health Administration (OSHA) recordkeeping rules in 1995 and 2002 may have led to substantial declines in the number of SOII recordable injuries and illnesses, particularly musculoskeletal disorders (Friedman 2007).

The causes of underreporting of nonfatal injuries and illnesses are complex and include both employer and employee factors. Employers may ignore or simply lack knowledge of recordkeeping requirements, minimize on-the-job injuries to maintain management bonuses, control workers' compensation insurance rates by direct payment for injuries, or fail to report in order to avoid targeted OSHA inspections or maintain eligibility for contracts requiring a good safety record. Employees may fear disciplinary action, lack knowledge of injury reporting requirements, become frustrated with workers' compensation procedures, or conform to peer pressure under incentive programs that reward a perfect safety record (Azaroff 2002). Based on these studies and others, a recent U.S. Congressional report in 2008 suggested that work-related injuries and illnesses in the U.S. are chronically and substantially underreported (U.S. House of Representatives 2008).

General Approaches to Occupational Injury and Disease Surveillance in California

The California Department of Public Health (CDPH) has conducted epidemiologic surveillance of work-related injuries and illnesses since 1984, when the legislature established a program to track the causes of workplace morbidity and mortality and conduct worksite investigations to generate prevention and intervention recommendations (California Health and Safety Code Section 105175). Under this mandate, CDPH has established surveillance systems to track selected work-related injuries and illnesses, including tuberculosis, lead poisoning, silicosis, carpal tunnel syndrome, needle stick injuries, pesticide-related illness, construction falls, asthma, traumatic fatalities, and heat-related illness. Based on this experience, we have developed and refined methods to ascertain, evaluate, and perform case and worksite investigations of acute injuries and illnesses, cumulative trauma, and diseases of long latency periods. Our specific approach to epidemiologic surveillance has varied by each health effect or hazard, but generally has followed several key principles:

• Multiple sources of surveillance data are preferred where possible to elucidate the causes and nature of workplace injury and illness.

• Passive surveillance methods are more advantageous than active surveillance systems.

• Cost-effective and timely systems are necessary to rapidly ascertain sentinel injuries and illnesses, identify sites for workplace investigations, and support the development and dissemination of prevention recommendations.

In particular, the California asthma surveillance program has modified the original sentinel provider-based reporting system to include multiple sources of case ascertainment, and has moved from active to passive case ascertainment. A survey of physicians we performed in 1993 indicated that less than 15% were willing to actively report cases directly to the SENSOR program (unpublished data). For work-related asthma surveillance, we elected to initially utilize an existing statewide reporting system (Doctors' First Reports of Occupational Injury and Illness) to increase case reports from throughout California and improve our efficiency in case ascertainment. We have since added emergency department, hospital discharge, and workers' compensation data as routine data sources for reporting.

Tracking Work-Related Asthma in California
With funding from NIOSH, CDPH has developed and maintained a multisource surveillance system for work-related asthma (WRA) in California since 1993. Asthma is a chronic respiratory disease of critical public health importance in the United States. The prevalence of asthma has been rising at an alarming rate, with a 75% increase of self-reported asthma between 1980 and 1994 (Mannino 1998) and a more moderate increase in recent years (NCHS 2009). Current estimates show that in 2004, there were approximately 21 million adults and nine million children with asthma in the U.S. (NCHS 2004). Asthma is associated with significant morbidity and economic costs. There were 13.6 million physician office visits, one million hospital outpatient visits, 1.8 million emergency department visits, and 3,780 deaths due to asthma nationwide in 2004 (NCHS 2004). In 2000, total costs due to asthma in the U.S. were estimated at $18.3 billion. This includes $10 billion in direct costs and $8 billion in indirect costs incurred by time lost from school, work, and premature deaths (AAFA 2000). Asthma is the fourth leading cause of work absenteeism, leading to nearly 15 million workdays lost each year (Mannino 2002). The Healthy People 2010 objectives call for the reduction of asthma deaths, hospitalizations, emergency room visits, and the number of school and work days lost; and an increase in the proportion of cases receiving asthma education and appropriate care (Objectives 24-1 through 24-7). The Centers for Disease Control and Prevention (CDC) also recommended surveillance for asthma in 25 or more states (Objective 24-8); this is now

Figure 1. Case Ascertainment Sources for WRA in California

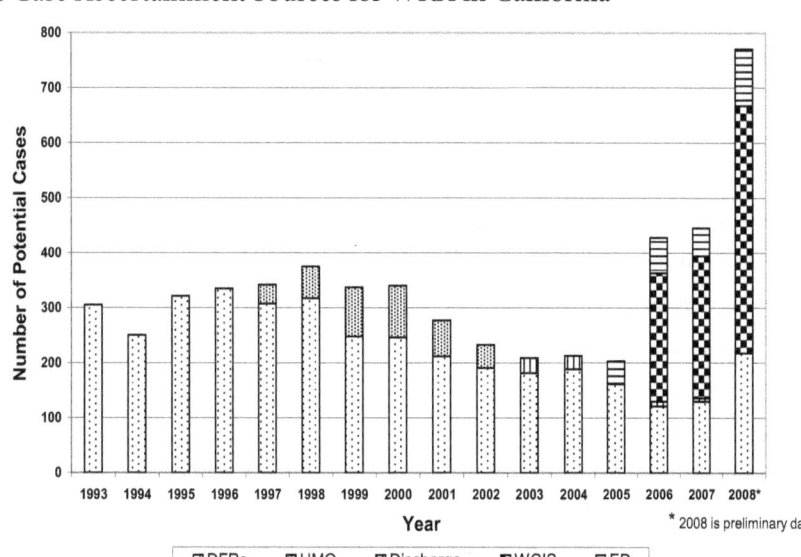

being implemented by the National Center for Environmental Health.

The same asthma trends have been documented in California. As the largest state, where one out of eight people in the U.S. lives and works, California bears a significant portion of the nation's asthma burden. The percent of adults ever diagnosed with asthma increased over the past 10 years, from 11.0 percent in 1995 to 13.7 percent in 2005 (Milet 2007). The California Health Interview Survey found that, in 2007, 13% of adults, or nearly 3.5 million people, have been diagnosed with asthma in the state (CHIS 2009). In 2005, these adults missed approximately two million days of work and over one-half million experienced daily or weekly asthma symptoms (Meng 2008). In 2005, Californians made 145,000 visits to emergency

Figure 2. Top 10 Occupations with Highest Rates of Work-related Asthma

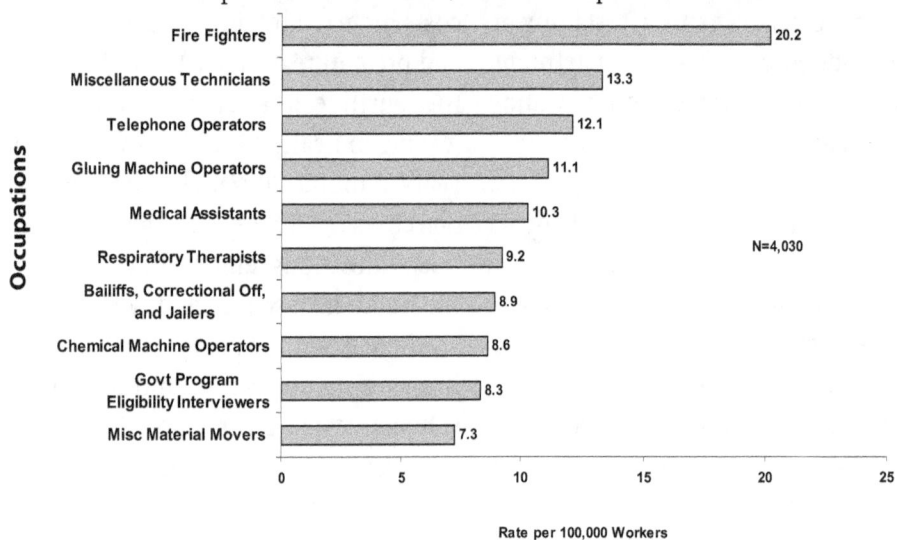

Figure 3. Top 10 Industries with Highest Rates of Work-related Asthma

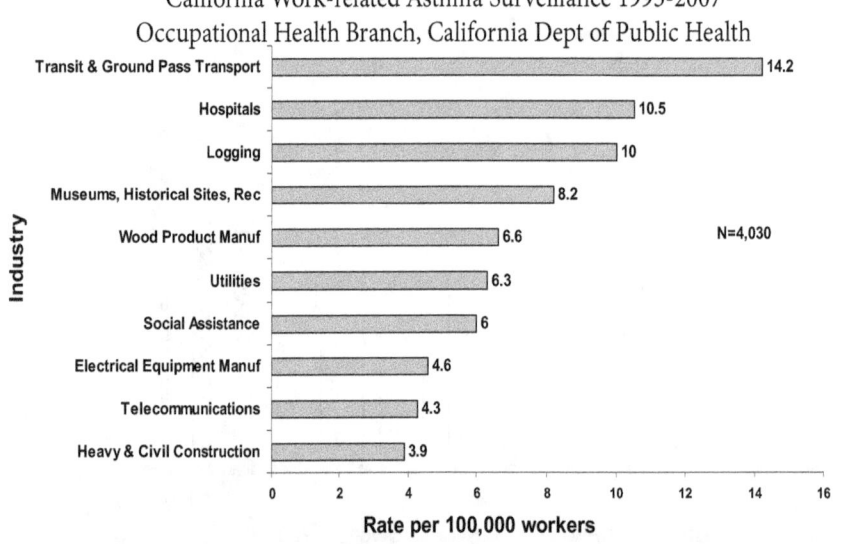

departments for their asthma, and were hospitalized 36,000 times. In 2004, 450 Californians died from asthma (Milet 2007). The economic impact of asthma in the state has been estimated to be over 2 billion dollars (AAFA 2009).

Work-related asthma has also been increasing (Meredith 1991, Chan Yeung 1994). It is estimated that workplace exposures are responsible for 15-20% of all asthma cases among adults in the U.S. (Balmes 2003, Toren 2009). In the industrialized world, WRA is now the most frequent non-asbestos occupational respiratory disorder (Toren 2000). The annual incidence rate for WRA ranges from 3 per million working people to 710 per million, depending on the country and study (Popin 2008, Orriols 2006, Vandenplas 2005, Mannino 2000, Jajosky 1999, Henneberger 1999, Karjalainen 2001, Milton 1998). Certain occupations have up to 30 times the average overall incidence rate in the U.K. (McDonald 2000). The prevalence of asthma in certain high-risk groups ranges from 5% in isocyanate and western red cedar workers (Vandenplas 1993, Chan Yeung 1993) to 9-10% in animal handlers (Meredith 1991) and firefighters (Ribeiro 2009), and up to 40-50% in detergent industry and platinum refinery workers (Venables 1987).

Figure 4. Cleaning agents used by janitor with WRA

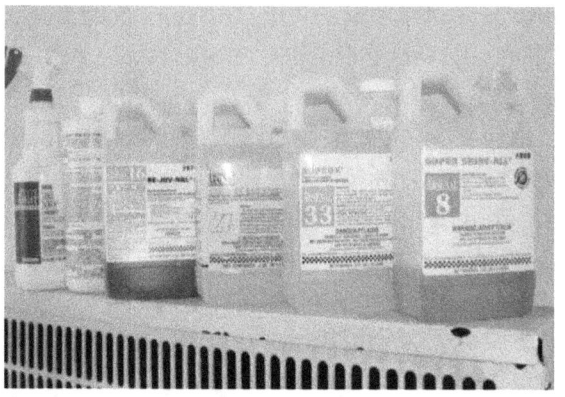

Figure 5. Janitor disinfecting bathroom with spray bottle

The California WRA surveillance program accounts for 53% of the cases in the national NIOSH data set (personal communication - Margaret Filios, NIOSH, 2009). The overall rate of WRA in California detected by the California WRA surveillance system (1.9 per 100,000) is similar to those identified by other surveillance systems in Michigan, the U.K., and South Africa (Reinisch 2001, Henneberger 1999, Hnizdo 2001, McDonald 2000). Underreporting is acknowledged to be an obstacle for all case-based WRA surveillance systems, resulting in a significant underestimate of individuals with this disorder (Esterhuizen 2001, Meyer 2001, Reinisch 2001). To overcome this problem, data from population-based surveys can be used to generate an estimate of the potential number of cases of WRA among the working population (Flattery 2006). Applying 2001 California Behavioral Risk Factor Surveillance System (BRFSS) data and the American Thoracic Society review estimate to year 2000 Census data, we estimate that 137,000 – 315,000 adults with asthma in California have asthma related to their work (Milet 2007).

Figure 6. Case detection for janitor with WRA due to cleaning agents

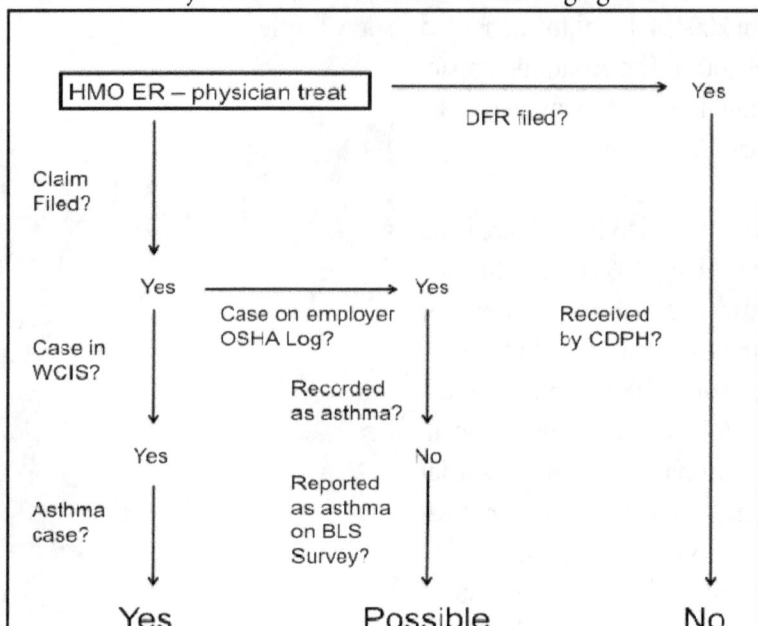

Over 15 years, the California WRA surveillance system has identified a total of 4,030 confirmed cases of WRA (annual average 269 confirmed cases). The overall rate of occupational asthma for all industries over 15 years is 1.9 per 100,000 employed in California. With the recent addition of three new statewide data sources (workers' compensation data, hospital discharge, and emergency department), the rate for 2007 has increased to 2.7 per 100,000 employed. The use of workers' compensation data (Workers' Compensation Information System – WCIS) has substantially increased case detection of WRA (Figure 1).

Industries with particularly high rates of WRA include local transit, hospitals, logging, museum and recreational sites, lumber and wood product manufacturing, utilities, social assistance, electrical equipment manufacturing, telecommunications, and heavy construction. Occupations with particularly high rates (Figure 2) include firefighters, technicians, telephone operators, gluing machine operators, medical assistants, respiratory therapists, correctional officers, chemical machine operators, welfare eligibility clerks, and material movers. Of the 4,030 confirmed cases with WRA from 1993-2007, 54% could be classified after interview and/or review of medical records. Of these, 7% were classified as Reactive Airways Dysfunction Syndrome, 10% as new onset asthma associated with a known asthma inducer, 41% as new onset cases associated with an unknown asthma inducer, and 43% as work-aggravated asthma.

Case Investigations and Followup for Work-Related Asthma

There are nearly 1 million health care workers employed in California. Hospitals represent the second highest WRA rate among specific industries in California (10.5 cases per 100,000 workers), with two of the top ten occupations (Figure 3) with the highest rates of WRA (medical assistants and respiratory therapists). WRA in health care industry workers is associated with exposure to a wide variety of agents including cleaning agents, formaldehyde, disinfectants, glutaraldehyde, and pesticides. A recent cluster of six cases of asthma has been reported to CDPH that are associated with the use of pre-packaged disinfectant wipes containing benzalkonium chlorides

(BACs). The BACs are listed as agents that cause asthma by the Association of Occupational and Environmental Clinics (AOEC 2009). Another case has recently been reported of a janitor in a large acute care hospital using a variety of cleaning agents, including the sensitizing agents BACs and diethanolamine. In this case, severe work-aggravated asthma occurred after this individual had used cleaning agents to disinfectant bathroom surfaces (Figures 4 and 5). This case illustrates the utility of a state-based multisource surveillance system to detect WRA, and identify causes of disease, and provide potential avenues for prevention and intervention.

The process of reporting the case described above illustrates how multi-source surveillance systems can improve identification of occupational illnesses (Figure 6).

After treatment in the emergency department of a health maintenence organization (HMO), a physician report was filed and sent to the employer's workers' compensation insurance carrier (Doctors First Report – DFR). Although the workers' compensation insurance carrier is required by law to submit a copy of the DFR report for tracking of work-related injuries and illnesses, the DFR in this case was not received by CDPH. This is consistent with a previous study that found that approximately one-third of all DFRs are captured by CDPH due to administrative problems with insurance carrier reporting. The workers' compensation claim was recorded by the employer on their electronic claims database management system, and also entered on the OSHA 300 log as a recordable injury and illness. As required by California law, the employer submitted a First Report of Occupational Injury (FROI) to the Workers' Compensation Information System (WCIS) that is administered by the California Division of Workers' Compensation (DWC). CDPH has legislative authority to utilize the WCIS to conduct surveillance of work-related injury and illness, and routinely requests WCIS data extractions for WRA under a Memorandum of Understanding with the DWC. This case was detected by the WCIS and verified as WRA according to our NIOSH surveillance case definition.

Although the BLS SOII conducts an annual nationwide sample of employers to collect detailed information about work-related injuries and illnesses, this system does not provide California with key information produced by the state-based WRAAR surveillance system, as explained below. If the employer was sampled as part of the BLS SOII, this case might be industry coded under general medical and surgical hospitals (NAICS code = 6221) and occupation coded under janitors and cleaners (SOC = 37-2011). Using the BLS 1992 Occupational Injury and Illness Classification System, this case might receive a Nature code for Extrinsic Asthma (1443), Part of Body code for Lungs (222), Source code for Cleaning and Polishing Agents (072), and Event code for Inhalation of Substance (341). However, as there were only 350 cases reported nationwide for Extrinsic Asthma in 2007, detailed data for WRA are not available for California alone. For example, in 2007 there were a total of 2,400 occupational illness cases reported by the California SOII for hospitals (NAICS code = 622), of which 200 were "respiratory conditions." Data on asthma cases from the California BLS SOII sample is not reported publicly due to insufficient sample size, and BLS is prohibited from sharing any identifying information that could be used for case follow-up, worksite investigations, and intervention studies.

Summary and Recommendations

State-based data sources that are not available on the national level can be used to identify specific cases and/or worksites for targeted investigations, thereby coordinating efforts at the individual and worksite level to reduce the burden of workplace injuries, diseases, and deaths. With the adoption of electronic data systems for hospital discharge, emergency room, ambulatory surgery, and workers' compensation claims, there is the opportunity to improve the ability of state and federal agencies to perform coordinated and timely surveillance that can more closely approximate the true nature and extent of workplace morbidity and mortality. While the BLS SOII is the most comprehensive nationwide sample of workplace injury and illness, it does not collect sufficient detailed data for many occupational illnesses and diseases that can be used to target worksites for interventions. Likewise, the BLS SOII system is not appropriate for ongoing and timely ascertainment of "sentinel" cases that link health outcomes to specific exposures that can lead to targeted worksite investigations and broad public health recommendations. State-based occupational health surveillance programs that can utilize the BLS SOII data in conjunction with workers' compensation and other public health data sets should be considered a key part of a comprehensive system to track occupational injuries and illnesses in the United States.

Beginning in 2009, the BLS has awarded funding to California, Washington, and Massachusetts to conduct a 3-year pilot study to enumerate selected workplace injuries and illnesses (amputations and carpal tunnel syndrome) with multiple data sources. These states will test the hypotheses that multi-source surveillance is needed to better elucidate the burden of work-related morbidity and mortality, and that utilization of existing electronic and other administrative data sets is a cost-effective mechanism to supplement the BLS SOII system. The long-term objective of these studies is to determine whether utilizing state-based workers' compensation and public health data systems for occupational safety and health surveillance is an effective strategy to improve our understanding of the causes and prevention of workplace injury and disease.

References

Association of Occupational and Environmental Clinics database of exposure codes; July 2009.

Asthma and Allergy Foundation of America. Costs of asthma in the U.S. 1994. Available: http://www.aafa.org/highcosts/index.html.

Azaroff LS, Levenstein C, Wegman DH. Occupational injury and illness surveillance: conceptual filters explain underreporting. Am J Pub Health [2002] 92:1421–1429.

Baker EL. Sentinel event notification system for occupational risks. Scand J Work Environ Health. 1988;14 Suppl 1:110-2.

Baker EL. Sentinel Event Notification System for Occupational Risks (SENSOR): the concept. Am J Public Health. 1989 Dec;79 Suppl:18-20.

Balmes JR, Becklake M, Blanc P, et al. Environmental and Occupational Health Assembly, American Thoracic Society. American Thoracic Society Statement: Occupational Contribution to the Burden of Airway Disease. Am J Respir Crit Care Med, 2003; 167:787-797.

Chan-Yeung M. Western red cedar and other wood dusts. In: Bernstein IL, Chan-Yeung M, Malo JL, Bernstein DI, eds. New York, Marcel Kekker Inc., 1993; pp. 503-531.

Chan-Yeung M, Malo JL. Aetiological agents in occupational asthma. Eur Respir J, 1994; 7:346-371.

Davis L: Role of State and Local Health Departments. In Levy BS, Wagner GR, Rest KM: eds., Preventing occupational disease and injury. American Public Health Association, 2005.

Esterhuizen TM, Hnizdo E, Rees D. Occurrence and causes of occupational asthma in South Africa-results from SORDSA's occupational asthma registry, 1997-1999, Sam J, 2001; 91:509-513.

Flattery J et al. The proportion of self-reported asthma associated with work in three states: California, Massachusetts, and Michigan, 2001. J Asthma 2006; 43:213-218.

Friedman LS, Forst L: The impact of OSHA recordkeeping regulation changes on occupational injury and illness trends in the U.S.: a time-series analysis. Occup Environ Med, 2007 64:454–460.

Henneberger PK, Kreiss K, Rosenman KD et al. An evaluation of the incidence of work-related asthma in the United States. Int J Occup Environ Health, 1999; 5:1-8.

Hnizdo E, Esterhuizen TM, Rees D, Lalloo UG. Occupational asthma as identified by the surveillance of work-related and occupational respiratory diseases programme in South Africa. Clin and Exp Allergy, 2001; 31:32-39.

Leigh JP, Marcin JP, Miller TR: An estimate of the U.S. Government's undercount of nonfatal occupational injuries. J Occup Environ Med, 2004 46:10–18.

Mannino DM. Occupational medicine: State of the Art Reviews, 2000; 15:359-368.

Mannino DM, Homa DM, Akinbami LJ et al. Surveillance for asthma-United States, 1980-1999. In: Surveillance Summaries, March 29, 2002. MMWR 2002; 51(No. 22-1):1-13.

Matte TD, Baker EL, Honchar PA. The selection and definition of targeted work-related conditions for surveillance under SENSOR. Am J Public Health. 1989 Dec;79 Suppl:21-5.

McDonald JC, Keynes HL, Meredith SK. Reported incidence of occupational asthma in the United Kingdom, 1989-97. Occup Environ Med, 2000; 57:823-829.

Meng YY, Babey SH, Hastert TA, Lombardi C and Brown ER. Uncontrolled Asthma Means Missed Work and School and Emergency Department Visits for Many Californians. Los Angeles: UCLA Center for Health Policy Research, 2008.

Meredith SK, Taylor VM, McDonald JC. Occupational respiratory disease in the UK 1989: A report to the British Thoracic Society and the Society of Occupational Medicine by the SWORD project group. Br J Ind Med, 1991; 48:292-8.

Meyer JD, Holt DL, Chen Y et al. Sword '99: Surveillance of work-related and occupational respiratory diseases in the United Kingdom. Occup Med, 2001; 51:204-208.

Milet M, Tran S, Eatherton M, Flattery J, Kreutzer R. "The Burden of Asthma in California: A Surveillance Report." Richmond, CA: California Department of Health Services, Environmental Health Investigations Branch, June 2007.

NCHS web site (http://www.cdc.gov/nchs/fastats/asthma.htm); Summary Health Statistics, National Health Interview Survey (NHIS), 2004.

NCHS (National Center for Health Statistics). 2009. Health data interactive. Accessed January 2009. http://www.cdc.gov/nchs/hdi.html.

National Institute for Occupational Safety and Health. Program Portolio – Surveillance. See http://www.cdc.gov/niosh/programs/surv/goals.html. Accessed January 2010.

Orriols R et al. Reported occupational respiratory diseases in Catalonia. Occup Environ Med 2006; 63:255-60.

Popin E, et al. The Incidence of occupational asthma in Alsace from 2001 to 2002. Results of intensification of the ONAP project in Alsace (2001-2002). Regional specificities. Rev Mal Respir 2008 Sep;25(7):806-13.

Rebeiro M et al. Prevalence and risk of asthma symptoms among firefighters in São Paulo, Brazil: a population-based study. Am J Ind Med 2009 Mar;52(3):261-9.

Reinisch F, Harrison RJ, Cussler S et al. Physician reports of work-related asthma in California, 1993-1996. Amer J Ind Med, 2001; 39:72-83.

Rosenman KD, Kalush A, Reilly MJ, Gardiner JC, Reeves M, Luo Z: How much work-related injury and illness is missed by the current national surveillance system? J Occup Environ Med 48:357–365, 2006.

Rutstein DD, Mullan RJ, Frazier TM, Halperin WE, Melius JM, Sestito JP. Sentinel Health Events (occupational): a basis for physician recognition and public health surveillance. Am J Public Health. 1983 Sep;73(9):1054-62.

Rutstein DD. The principle of the sentinel health event and its application to the occupational diseases. Arch Environ Health. 1984 May-Jun;39(3):158.

Stanbury M, et al: Guidelines for Minimum and Comprehensive State-based Public Health Activities in Occupational Safety and Health. DHHS (NIOSH) Publication No. 2008-148.

Toren K, Brisman J, Olin AC, Blanc PD. Asthma on the job: work-related factors in new-onset asthma and in exacerbation of pre-existing asthma. Respir Med, 2000;94(6):529-35.

Toren K, Blanc P. Asthma caused by occupational exposures is common—A systematic analysis of estimates of the population-attributable fraction. BMC Pulmonary Medicine 2009, 9:7.

US House of Representatives Majority Staff Report by the Committee on Education and Labor. Hidden Tragedy: Underreporting of Workplace Injuries and Illnesses, June 2008. See http://edlabor.house.gov/labor/worker-safety-and-health/.

Vandenplas et al. Epidemiologie de l'asthma professional en Belguique. Rev mal Respir 2005; 22:421-30.

Venables KM. Epidemiology and the prevention of occupational asthma. Br J Ind Med, 1987; 44:73-75.

Managing Prevention with Leading and Lagging Indicators in the Workers' Compensation System

Benjamin C. Amick III, Institute for Work & Health, University of Texas School of Public Health, Sheilah Hogg-Johnson, Institute for Work & Health

It is not uncommon parlance to say 'what gets measured gets managed'. Workers' compensation systems have long used claims rates to try and improve the performance of employers and the workers' compensation system. Labour Departments, insurers and workers' compensation agencies use workers' compensation data to identify and target firms needing inspection. While often used to establish economic incentives – including insurance costs and rebates - the traditional information in a workers' compensation system does not include leading indicators commonly used in business for the management of organizational performance. In this paper, we present some recent work by the Ontario prevention system to develop leading indicators of organizational performance and finally, some collaborative research between the Institute for Work & Health (IWH) and Ontario occupational health and safety system partners.

Managing with Lagging Indicators: the Ontario High Risk Firm Initiative

The Ontario High Risk Firm Initiative was a program of targeted Occupational Health & Safety enforcement and consultation that ran from 2004 to 2008 with the aim of reducing the province wide rate of time-loss injuries from 2.2 per 100 workers to 1.8 per 100 workers. The initiative was intended to be an integrated prevention strategy involving the key prevention partners in Ontario: the Ministry of Labour (MOL) with responsibility for enforcement of the Occupational Health & Safety Act and regulation; twelve sector specific Health & Safety Associations (HSAs) providing consultation and education services to workplaces; and the Ontario Workplace Safety & Insurance Board (WSIB), providing compensation to injured workers.

In the High Risk Firm Initiative, all firms registered with the Ontario WSIB were ranked each year based on work injury statistics using a combination of indices such as work injury claim rates and costs. The 10% of firms with the worst rankings were to be targeted each year. The worst 2% were targeted with enforcement and enhanced inspections – four per year. The next 8% were called "Last Chance" firms in the program and were referred to the appropriate HSAs for targeted consultation and education. If an HSA could not target all firms on their lists, then some firms could be referred back the MOL for a priority inspection – a single enhanced inspection in the year. In order to launch this initiative, the MOL hired additional inspectors to increase the number s from 3.8 to 5.4 per 100,000 workers.

The MOL has declared the High Risk Firm Initiative a success as the goal of reducing the lost-time claim rate from 2.2 to 1.8 per 100 workers has been achieved. Each of the four High Risk Cohorts shows declining lost-time claim rates concurrent with the introduction of the intervention. Evaluation of an initiative such as this can prove challenging, though. Because all firms within the worst 2% were targeted, there is no suitable concurrent control group available for comparison, meaning that one must rely upon either before/after designs, or on historical controls. In addition, whenever interventions are targeted based on extreme values of a measure, there is the threat of regression to the mean.

Figure 1. Conceptual Model for Primary and Secondary Prevention Integrated with Injuries, Illness and Work Disability Outcomes.

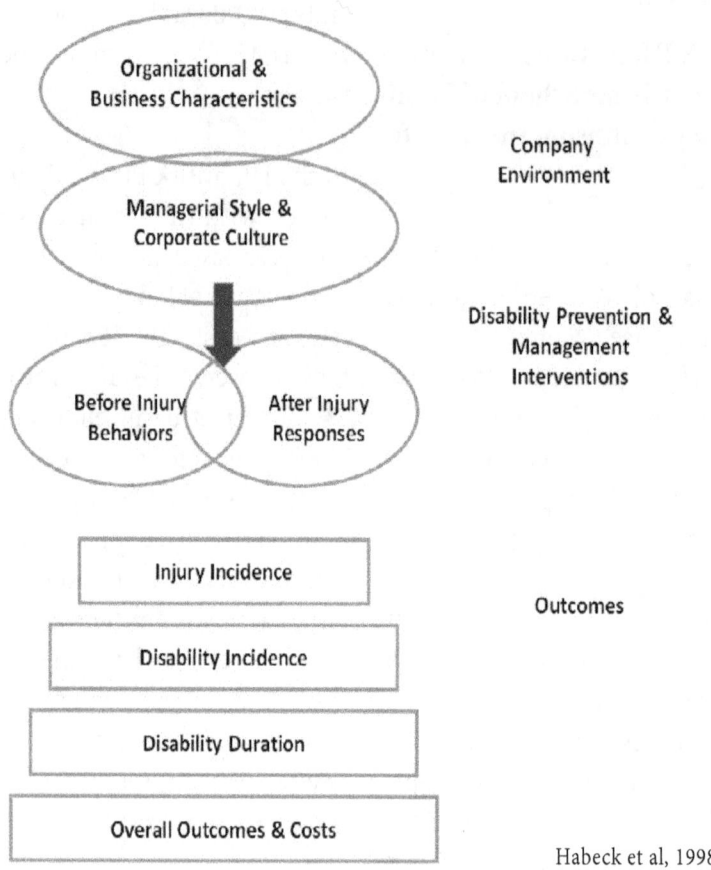

Habeck et al, 1998

Is There an Evidence Base for Leading Indicators?

Brewer (2007) recently examined the evidence on the effectiveness of injury and illness prevention and work disability management programs in injury prevention and loss control. The only strong evidence indicates disability management/return to work programs effectively reduce lost time claims and associated costs. Research was mostly clinically-based, calling attention to graded activity (Stahl, 2004), rehabilitation (Jensen, 2005), therapy (Durand, 2001; Feuerstein, 1993), early intervention (Greenwood, 1990) and disability case management (Arnetz, 2003; Loisel; 2002). Only one study focused on changing RTW policies (Brown, 1992). These findings support a view of prevention that integrates primary prevention with secondary prevention (Frank and Cullen, 2005).

A recent examination of systematic reviews on the effectiveness of workplace interventions to reduce musculoskeletal injuries identified the importance of worker participation in prevention programs (Amick 2009). There were few studies, however, that examined the influence of organizational management participation or organizational management programs (e.g., safety leadership) on worker health outcomes. Another recent scoping review to determine if there was enough evidence to complete a full systematic review, again by Brewer (2006), examined the evidence for safety culture or safety climate programs that could demonstrate an effect on injuries or illnesses. The research team concluded there was not enough published evidence to support a full review and very little evidence to support either culture or climate initiatives. Finally, Robson

> Questions for Measures of Organizational Performance
>
> 1. Formal safety audits at regular intervals are a normal part of our business.
> 2. Everyone at this organization values ongoing safety improvement in this organization.
> 3. This organization considers safety at least as important as production and quality in the way work is done.
> 4. Workers and supervisors have the information they need to work safely.
> 5. Employees are always involved in decisions affecting their health and safety.
> 6. Those in charge of safety have the authority to make the changes they have identified as necessary.
> 7. Those who act safely receive positive recognition.
> 8. Everyone has the tools and/or equipment they need to complete their work safely.

(2005) examined the effectiveness of occupational health and safety management system (OHSMs), and Bigelow (2005) examined the literature for OHSMs audit tools. Both found little evidence to support effectiveness or recommend tools. We are left with an interesting conclusion -- that there is limited evidence on effective policies, programs, and practices at the organizational and management level to suggest new evidence-based leading indicators. Clearly, there is a need for new approaches and new tools for measuring organizational and management policies, programs and practices important in predicting injuries, illnesses and work disability.

As a first step, it may be appropriate to identify a conceptual model or schemata that easily integrates primary and secondary prevention, but clearly describes how these activities are to affect injuries, illness and work disability. Hunt (1993) and Habeck (1998) have proposed a relatively straightforward model (Figure 1). In this model, features of the business interact with management style and culture to affect prevention programs (both pre- and post-injury). These, then, have an effect on outcomes. While simple in conceptualization, early on the model recognized the importance of integrating primary and secondary prevention programs and policies and looking at the broader functioning of the organization of not just being siloed in occupational and environmental health programs. It suggested both multiple predictors of outcome and multiple outcomes, allowing for either complex or simple relationships to emerge. The Hunt and Habeck model provides a useful series of organizing principles for the consideration of a set of organizational and management measures and a clear set of outcomes these measures would be expected to predict.

Developing Leading Indicators in Ontario

In Ontario, MOL, WSIB and sectorally-based HSAs along with labour and labour-based organizations have responsibility for occupational safety and health. The Occupational Health and Safety Council of Ontario (OHSCO) serves as an umbrella organization to integrate the work of each of these important prevention partners. OHSCO asked the IWH to collaborate with all prevention system partners to develop a measure of safety culture that was short and easy to use by health and safety consultants (front-line service staff working with employers). The group felt measuring safety culture was inappropriate for a consultant but did agree on a series of questions about organizational performance. A consensus was reached on 8 questions shown above. The

8 questions are currently being piloted and, if found reliable and valid, then all consultants in the prevention system will be asked to obtain this information from all contacts. The information will be warehoused at IWH where it will be linked to lagging indications (claims rates) and reports will be generated by sector for use in managing the system. This work has led to consensus in the province that developing a series of valid and reliable leading indicator measures is important.

Benchmarking Leading Indicators in Ontario – A Collaboration

The challenge with developing benchmarking initiatives for leading indicators is the lack of a strong evidence base. The long-term objective of the collaboration is to build a scientifically-grounded evidence base for a generalizable set of leading indicators to benchmark how firms organize and manage occupational health and safety (OHS), which are relevant to all sectors and firms in Ontario. Three specific objectives are: 1) to identify a reliable and valid set of firm-level metrics of organizational and management behaviour relevant to OHS and usable by the OHS community; 2) to examine the relationships between WSIB claim rates, claim durations and organizational and management metrics; and 3) to demonstrate a scientifically-grounded procedure for collecting valid firm-level organizational metrics, aggregating the data and disseminating benchmarking information to key stakeholders in Ontario.

To accomplish this we will be surveying 5,000 firms in 5 sectors (service, manufacturing, electrical and utilities, pulp and paper, municipal) to assess:

• Organizational policies and practices in 1° and 2° prevention following Amick (2000)

• Safety culture

• OHS management systems following Fernandez-Muniz (2006, 2007)

• Employee relations/joint health and safety committee functioning following Shannon (1996) and Geldhart (2006)

Little scientific evidence exists to establish the best methods for producing unbiased benchmarking data and useful benchmarking reports. The proposed research will address how to produce unbiased and useful information. We will examine who is the best person to talk with in the organization about occupational health and safety information. We will consider whether some businesses are systematically underrepresented or overrepresented and propose approaches for making credible statements about industry benchmarks. Finally, we will work with key stakeholders to define a useful benchmarking report.

Summary

Ontario province has begun to examine the benefits of using leading indicators in combination with lagging indicators to manage the performance of occupational health and safety systems.

References

Amick III BC, Habeck RV, Hunt A, Fossel AH, Chapin A, Keller RB, Katz JN. Measuring the impact of organizational behaviors on work disability prevention and management. J Occup Rehabil 2000;10(1):21-38.

Arnetz BB, Sjogren B, Rydehn B, Meisel R. Early workplace intervention for employees with musculoskeletal-related absenteeism: A prospective controlled intervention study. J Occup Environ Med 2003;45(5):499-506.

Bigelow PL, Robson LS. Occupational health and safety management audit instruments: A literature review. Toronto: Institute for Work & Health; 2005.

Brewer S, King E, Amick III BC, Delclos G, Spear J, Irvin E, Mahood Q, Lee L, Lewis C, Tetrick L, Gimeno D, Williams R. A systematic review of injury/illness prevention and loss control programs (IPC). Toronto: Institute for Work and Health; 2007.

Brewer S, King E, Delclos G, Spear J, Irvin E, Mahood Q, Lee L, Lewis C, Tetrick L, Williams R, Amick III BC. Scoping review of the sustainability of safety culture/climate in the workplace. Toronto: Institute for Work and Health; 2006.

Durand MJ, Loisel P. Therapeutic return to work: Rehabilitation in the workplace. Work 2001;17(1): 57-63.

Fernandez-Muniz, B, Montes-Peon, JM, Vazquez-Ordas, CJ. Safety management system: Development and validation of a multidimensional scale. Journal of Loss Prevention in the Process Industries 2007;20:52-68.

Fernandez-Muniz, B, Montes-Peon, JM, Vazquez-Ordas, CJ. The relationship between occupational safety management and firm performance. Saf Sci 2008; 45:1-2

Feuerstein M, Nicholas RA, Huang GD, Dimberg L, Ali D, Rogers H. Job stress management and ergonomic intervention for work-related upper extremity symptoms. Appl Ergon 2004;35(6):565-574.

Frank J, Cullen K. Preventing injury, illness and disability at work. Scand J work Environ Health 2006; 32(2):160-167.

Geldart S, Shannon HS, Lohfeld L. Have companies improved their health and safety approaches over the last decade? A longitudinal study. Am J Ind Med 2005; 47(3):227-236.

Greenwood JG, Wolf HJ, Pearson RJ, Woon CL, Posey P, Main CF. Early intervention in low back disability among coal miners in West Virginia: Negative findings. J Occup Med 1990;32(10):1047-1052.

Habeck RV, Hunt HA, VanTol B. Workplace factors associated with preventing and managing work disability. Rehabil Couns Bull 1998;42(2):98-143.

Hunt HA, Habeck RV, VanTol B, Scully, SM. Disability Prevention Among Michigan Employers. Final Report submitted to the Michigan Department of Labor (Upjohn Institute Technical Report No. 93-004). Kalamazoo, MI: W.E. Upjohn Institute for Employment Research; 1993.

Jensen IB, Bergstrom G, Ljungquist T, Bodi, L. A 3-year follow-up of a multidisciplinary rehabilitation programme for back and neck pain. Pain 2005;115(3)273-283.

Loisel P, Lemaire J, Poitras S, Durand MJ, Champagne F, Stock S, Diallo B, Tremblay C. Cost-benefit and cost-effectiveness analysis of a disability prevention model for back pain management: A six year follow up study. J Occup Environ Med 2002;59(12):807-815.

Shannon H, Walters V, Lewchuk W, Richarson J, Moran LA, Haines T, Verma D. Workplace organizational correlates of lost-time accident rates in manufacturing. AJIM 1996;29:258-268.

Staal JB, Hlobil H, Twisk JWR, Smid T, Koke AJA, van Mechelen W. Graded activity for low back pain in occupational health care: A randomized, controlled trial. Ann Intern Med 2004;140(2):77-84.

Benchmarking and Performance Measurement for Governments

Mary L. Stewart, ARM, CPCU, Public Entity Risk Institute (PERI)

The PERI Data Exchange Overview

The Public Entity Risk Institute (PERI) developed a unique data warehouse of detailed information about public entity liability and workers' compensation claims and exposures called the Data Exchange. The primary goal of the Data Exchange is to serve as a benchmarking tool for government managers and officials interested in comparing their claims experience with peers to improve their risk management programs, make better-informed decisions, and control their liability and workers' compensation costs. All the data in the Data Exchange is supplied voluntarily by participating public entities, public risk pools, and third party claims administrators. The program collects data elements that include financial information (such as the total amount paid for a claim) and dimensional data (such as the cause of loss and nature of injury). The data elements are then separated into one of two "data marts" created to track either liability or workers' compensation claim activity. This information is then translated into a series of free reports designed to promote benchmarking metrics and is made available to participants through the PERI website.

Data Exchange Participation

The following tables identify the demographics of the governments that are active in the Data Exchange. Exhibit A is a summary of the number of entities participating in the liability data mart (LB) and the workers' compensation data mart (WC) and Exhibit B highlights the participation in both data marts by type of government.

How the Data Exchange Works

By participating in the Data Exchange, governments can measure their performance against desired frequency and severity metrics and can link their activity to processes and programs that will target their improvement. Claims data for governmental entities is provided to PERI from various sources (i.e., third-party administrators, self-insureds, pool administrators, insurance carriers, and consultants). To ensure that the data is comparable, the Data Exchange maps each and every code from multiple data suppliers across a series of fields, such as cause of injury, cause of loss, part of body, type of vehicle, nature of injury, event and exposure, employee positions, and department. This allows PERI to collect comparable data without suppliers changing their internal coding systems.

Peer Groups

A peer group is defined as a subdivision of governments based on similar size or a demographic, such as population, payroll, number of employees, miles driven, average daily attendance for schools and higher education, operating budget, etc. PERI divided public organizations into one of the following government types: Cities and Towns, Counties, Federal, Higher Education, Housing Authorities, Indian Reservations, Public School Districts, Public Transits, Special Districts, and State. Combining the type of government with the geographical location added validity to the comparisons. For example, a large transit system or an airport may not have a comparable peer in their state, but can locate several comparable peers within their region (NOTE: PERI uses the four regions created by the US Postal System).

Exhibit C identifies a sample of the different size bands used to compare different governments based on their annual payroll, number of full-time equivalent employees, and population. A town with less than 500 employees surrounded

Exhibit A. Participation and Claim Summary (as of 8/15/2009)

Data Mart	No. Entities	Claim Count	Claim Value
WC Total	4,967	679,347	$6,105,115,162
LB Total	4,034	200,361	$1,276,049,805

Exhibit B. Participation by Type of Government (as of 8/15/2009)

	WC Data	LB Data
Cities & Towns	1,191	1,146
Counties	363	250
Higher Education	46	5
Housing Authority	158	157
Pool Administration	20	16
Public School Districts	599	393
Public Transit	23	43
Special Districts	2,566	2,024
State Government	1	-

Exhibit C. Sample Peer Group Bands

Payroll	Employees (FTE)	Population
Less than 5,000	Less than 500	Population under 1,000
5,001-15,000	501-4,999	Population 1,001 to 10,000
15,001-25,000	5,000-13,999	Population 10,001 to 25,000
25,001-50,000	14,000-22,999	Population 25,001 to 50,000
50,001-100,000	23,000-31,999	Population 50,001 to 100,000
100,001-250,000	32,000-40,999	Population 100,001 to 300,000
250,001-500,000	41,000-49,999	Population 300,001 to 500,000
500,001-1,000,000	50,000-59,999	Population of 500,000 or more
Over 1,000,000	Over 60,000	

by large counties needs to find other peers with similar exposures in their state or within the same region to adequately compare statistics for benchmarking.

Some of the government types are further divided into sub-divisions to increase their ability to find best practices within unique government types. For example, special districts are further divided by type of service: abatement districts, aviation districts, cemetery districts, community service districts, conservation districts, emergency districts, legal districts, library districts, medical districts, planning/economic districts, port authorities, recreation districts, sanitary districts, telecommunication, transportation districts, utility districts, and water districts. Each of the divisions actually represent a group of governmental units located anywhere in the United States that have some similarity and yet are uniquely different depending on the federal, state and local requirements.

Special districts represent the fastest growing type of government, yet in many states the claim activity is not truly identified against the exposure potential within a given district. The difficulty is trying to relate common activities (i.e., flood control, water diversion, quality testing) and loss exposures (i.e., infrastructure deterioration, chemical usage, equipment operation, acts of God) among the same type of districts without considering all of the different variables (i.e., funding, policies, procedures, geographical conditions, political involvement, cooperative arrangements) each organization has. When different variables are analyzed against common activities comparison becomes impossible.

Reports
The data itself is confidential; only those organizations submitting claims data have permission to view the results in web-based reports. Since the reports are available on-line at any time of day, a benchmark seen at the beginning of a month may change when the same report is pulled at a later date as more data is added to the Data Exchange. Each Data Mart offers a particular library of reports using the metrics defined by PERI. Information is usually displayed in at least three ways: claim cost in dollars, percentage by claim cost, and percentage by claim count.

Benchmarking Results
Benchmarking helps to incorporate best practices into your operation by finding other organizations with similar exposures and policies that have better results or established standards. Benchmarking against other governments reduces uncertainty; provides objective, quantifiable documentation of results; increases credibility to claim evaluations; and is a forecasting tool to estimate loss experience.

Comparison Limitations
PERI has been working on this project for over six years and has discovered several issues that continue to limit the success of claim comparison. Some of the challenges that need to be fixed before claims analysis can improve include:

Lack of Standards
Public entities, risk pools and third party administrators have invested in proprietary coding systems that meet their internal needs, but do not enable comparison for benchmarking purposes. Many governments are tracking heart related claims covered under workers' compensation because of state laws, and yet those claims could be classified as "heart," "cardiovascular," or "heart/lung related". Also, when one client switches vendors, their past claims data must be recoded to match the new claims system. The lack of industry standards limits the recognition of similarities and differences, which could provide useful information to prevent future claims.

Too Detailed to Find Related Events

One jurisdiction used over 9,000 possible choices under "cause of loss" and then wondered why they could not find significant comparisons. It would be better if more organizations used a tree approach to delineate similar starting points. For example, PERI started the project ready to map over 300 fields of workers' compensation information and quickly found most organizations were unable to provide that level of detail. Currently, PERI requests 68 different fields and consistently receives about half from the data suppliers.

Processing Errors

"Cleaning" the data is a full time job. Insufficient quality control has become a major issue that reduces the reliability of the data collected. PERI received more than 1.5 million claims over six years for accidents occurring between 1999 and 2009. Some of the common errors should have been corrected by the clients, their broker, or the third-party administrator (i.e., multiple names used for the same client; reported dates happening before the date of loss; closed claims with reserves; and a high level of missing information in a coded field). While the issue was once centered on the difficulty of finding the errors, now the greater issues rests with the ever increasing attitude it's not worth the effort to fix them.

RMIS System Not Designed for Analysis

Most Risk Management Information systems (RMIS) were designed to pay claims; the systems were not designed to store loss control or litigation information that might be useful for another source. In fact, many of the public entities indicated they are charged extra when the claim adjusters load additional claim facts into the third-party system. Once again, the data may be lost if it is not stored in an electronic system and maintained over several years. Too many governments loose the electronic copy of the claims data when they switch claim adjusters, and are left with a paper version that is out of date.

Reduction of Manual Files

The insurance industry has been trying to eliminate paper files, and yet many of their clients continue to keep paper files, which contain claim details in various records from internal investigations, outside consultants, and medical reports. When the data is loaded into a RMIS system, some of the information is still hard coded, which is impossible to use in any electronic sorting system. Claim tracking practices need to eliminate the inefficiency of manual entries and adopt an electronic approach to compare all information related to the claim.

Shared Goals

While political figures have "pet" projects, all organizations have set goals that support their mission and corporate culture. Unfortunately, the collection of historical data is not high on the list; even when the analysis of those statistics could reduce the frequency and severity of employee accidents. Until more resources are directed to improve the collection of data, the issues mentioned above will continue to limit what could be learned from each others' experiences.

About PERI

The Public Entity Risk Institute is a nonprofit, non-membership organization that provides risk management education and training resources for local governments, school districts, small businesses, nonprofits and others. Its website serves as a clearinghouse and library with information on a wide range of topics including disaster management and hazard mitigation, environmental liability, risk financing and insurance, education, safety and health protection, workers' compensation and technology risks. PERI provides resources that include publications, training, internet symposia, tools, data, and data analysis. More information about PERI and the Data Exchange can be found at www.riskinstitute.org.

Self-Insured Experience with Workers' Compensation

Robert B. Steggert, Marriott International, Inc.

Overview

Marriott International is self-insured and self-administered in the United States. Claims administration is augmented by various in-house, nurse-centric programs designed to assure timely, appropriate and quality care. Financially, Marriott utilizes a "carrot-and-stick" profit and loss chargeback system designed to drive management ownership and proactive loss prevention. Loss reduction pressure may create unintended consequences, e.g. basic on-time reporting and transitional RTW coordination.

Core loss prevention programs include:
- On-time reporting focus
- Loss rate communication
- Risk Ranking
- Accident analysis by department and type

Data in the tables below are percentages of 5,763 accidents with a total loss of $17,862,309 in 2008.

Table 1. Marriott Lodging Workers' Compensation Accidents by Type (2008)

Type	% of Accidents	% of Losses
Struck by Object	23%	19%
Slip & Fall	20%	26%
Weight a Factor	14%	36%
Cut	14%	4%
Burn	7%	4%
Chemical	2%	3%
Repetitive Motion	2%	4%
Object in Eye	2%	2%
Unknown	7%	<1%
Other	8%	4%

Table 2. Marriott Lodging Workers' Compensation Accidents by Department (2008).

Department	% of Accidents	% Losses
Housekeeping	38%	43%
Kitchen	16%	14%
Banquets	11%	12%
Restaurant	7%	6%
Engineering	6%	9%
Front Office	5%	3%
Administration	6%	5%
Laundry	3%	5%
Recreation	2%	2%
Other	4%	

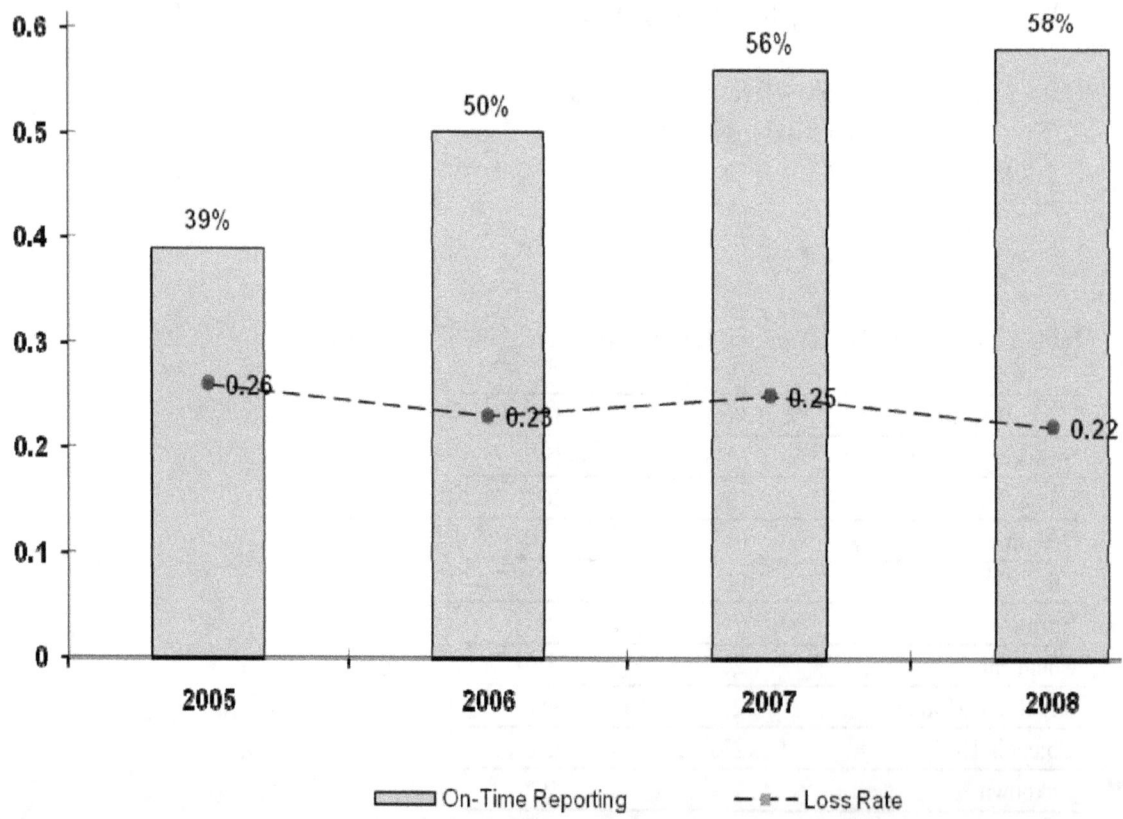

Figure 1. Proportion Reported on Day of Event and Loss Rate, 2005 – 2008

Using Workers' Compensation Data: The Move from Lagging to Leading Indicators

Sandra Carson, Sysco

Workers' compensation data can be used in programs to prevent workplace injuries and illnesses. However, to obtain useful data, employers must require workers' compensation carriers and/or third party administrators to collect information beyond the basic data traditionally collected by the insurance industry. Historically, the gathering of data by these groups was done primarily to set up claims and submit required reports. Use of workers' compensation data inherently has the following issues:

- information is primarily insurance industry driven "claims information"

- variable accuracy depending on the persons collecting and inputting the data

- competing goals (i.e. Insurance carrier gross revenues actually increase with claim costs, TPAs are paid by claim, claims management is measured by handling claims to closure not preventing them in the first place, etc.)

- most information gathered is single dimension (linear) not in the context of other factors

- usually doesn't account for findings/conclusions from an accident investigation

- lacks many data sets needed to do a thorough analysis

- groups who purchase the insurance (risk management) or selects the TPA (claims) are not tuned in to the needs of occupational health/safety where injury prevention is driven

This case study is with a fortune 100 company with over 150 locations employing over 45,000 employees in the United States. Because the company has historically been decentralized, fostering an entrepreneurial environment, each site has a separate employer with its own Chief Executive Officer/President. This lent itself to flexibility at a local level with innovative plans form fit for the environment. While there have been many realized benefits from this de-centralization, there are down-sides including a lack of benefiting from economies of scale. It has also resulted in a lack of consistency in the data collected and processes associated with occupational illness/injury prevention and failure to benefit from learning from the other locations' data.

Occupational health/safety and workers' compensation claim handling were identified as areas which would benefit from centralization and corporate oversight. Many different options were considered for centralizing injury and illness data, including separate systems for occupational injury/illness data and workers' compensation. It was determined that improving and broadening the data collected by the claims system would be the easiest strategy, least costly and held the best potential for success as a step to better tap the potential of this information for injury prevention.

The use of workers' compensation information for this company is as an "alarm activation" or trigger system pointing efforts toward a systemic injury symptom allowing for triage on a macro level. It is only one resource and certainly not the pinnacle or sole source. To benefit from this centralization of data for analysis, many hurdles had to be addressed such as:

- specific data fields needed are not available in the current system

- data fields added would not have historical context

- persons inputting the information did not focus on accuracy just speed/ease of input

- generic fields such as "other" or "not otherwise specified" were improperly used

- output reports didn't capture additional fields necessary to conduct a meaningful analysis

The result of addressing these issues is the ability to "pro-act" to objective information collected through the workers' compensation claim data leading to a reduction of occupational injuries/illnesses and the related costs. Specific examples will be discussed in this session. While this company is placing a heavy emphasis on moving from lagging to leading indicators, the use of capturing workers' compensation data as an indicator detection system will continue.

Past, Present, and Future Uses of Some Workers' Compensation Data

J. Paul Leigh, Center for Healthcare Policy and Research, University of California, Davis

Introduction

This paper will address some of the workers' compensation studies – past, present, and future – with which I am familiar. I will also offer some comments on data sources. This paper is not my view of the "most important" studies and data. There are three messages. First, workers' compensation does not get the research attention it deserves from physicians, nurses, epidemiologists, and social scientists. Second, there are great varieties of studies and those varieties are growing; workers' compensation studies are not limited to only a handful of questions. Third, some data sources are more useful than others.

Workers' Compensation Deserves More Research Attention

National workers' compensation costs have been estimated at roughly $87.6 billion for 2006 (1). Liberty Mutual, one of the largest workers' compensation insurance carriers, estimated direct and indirect costs for all occupational injuries and illnesses (not just those covered by workers' compensation) to be from $155 billion to $232 billion in 1998 (2). These costs are large when compared to those for other diseases. The costs of cancer in 2005 were estimated to be $210 billion (3). The costs of diabetes in 2007 were estimated to be $174 billion (4). A recent estimate of the costs of Alzheimer's disease was $148 billion (5).

Liberty Mutual's cost estimates indicate a higher percentage of indirect costs (e.g., lost wages) to total costs when compared to other diseases. The reason is that over 70% of these costs are due to injuries rather than illnesses and injuries and account for more harm to younger persons than are accounted for by diseases. Occupational injury deaths, for example, frequently occur among persons in their 30s and 40s whereas cancer and especially heart disease and Alzheimer's deaths frequently occur among persons in their 70s and 80s. Whereas all deaths are losses, deaths among younger persons who may have young children are especially tragic.

Yet, as anyone who watches TV or listens to the radio or reads newspapers will tell you, cancer, diabetes, and Alzheimer's are mentioned far more often than workers' compensation or occupational injury and illness.

These cost estimates invite comparisons to federal government funding for health research. The National Institute for Occupational Safety and Health (NIOSH) has consistently received among the smallest amounts of funding compared to other institutes. The 2006 fiscal year funding for NIOSH was $254 million. This compares to: $4,793 billion (19 times NIOSH budget) for the National Cancer Institute; $1,844 billion (over 7 times NIOSH) for the National Institute for Diabetes, Digestive and Kidney Disorders; and $508 million (2 times) for the National Institute of Arthritis and Musculoskeletal Disorders (6).

Examples of the Variety of Past, Present, and Future Research

Past and Present. From an economics perspective, one of the oldest research questions involves the extent to which the ratio of cash benefits to wages (replacement ratio) for workers encourage filing claims.(7-9) Many studies suggest that a 10% increase in benefits is associated with a 1% to 4 % increase in the number of claims and duration of claims. But methodological problems afflict most studies. For example, few studies with which I am aware account for the fact that state legislators may increase or decrease benefits due in part to the number of workers recently injured (and who might vote) in the state or due to rapidly escalating workers' compensation insurance premiums among carriers in the state.

The wage-replacement ratio has also been researched from the perspective of equity. A recent summary of ratios for permanent partial disability (the workers' compensation category generating the greatest total costs) suggested a range of 29% to 46% (10). But these authors pointed out that this range applied to the late 1980s and early 1990s and that legislated benefits have decreased since that time. These authors conclude that the actual ratios are considerably less than the 2/3rds ratio cited by others as the most reasonable of equity standards. A different study on equity considered the changes in benefits that occurred after the historic 2004 change in California law. Estimates showed an average reduction of more than 50% (11). Studies by researchers paid by insurance companies and the California Commission on Health and Safety and Workers' Compensation (CHSWC) later verified a roughly 50% reduction (12, 13).

A related area of labor-market research involves the "compensating wage hypothesis." This hypothesis holds that in competitive labor markets, workers in dangerous jobs are paid a higher wage than similar workers in safe jobs; the higher wage compensates them for taking the dangerous job. Debate surrounds the extent to which labor markets actually pay such a wage, but all economists would agree that markets would be more likely to pay this wage if information on job hazards were widespread. But there is some question about whether or not this information is widespread. One difficulty in measuring hazards is that they are so dissimilar. Carpal tunnel syndrome is different from back strain which, in turn, is different from a fatality due to a vehicle crash. Workers' compensation costs might be viewed as an aggregate summary of these various injuries and diseases – as a summary of aggregate job risks. Moreover, this summary is measured with a metric that is easily understood by most Americans: dollars. The provision of information on costs, from this perspective, should improve market efficiency, provided that costs are associated with specific occupations, industries, and jobs. Some studies have attempted to provide cost-per-employed-worker (not cost-per-injury) across occupations and industries (14, 15).

Another line-of-research investigates whether and how much workers who would likely quality for workers' compensation do not apply for workers' compensation benefits. One study estimates over 40% never apply (16). As reported elsewhere, Leigh (17) and Lakdawalla et al (18) find that the "more vulnerable workers" without any other health insurance are more likely than affluent workers to never file for workers' compensation. The Gilmore Research Group (19), in cooperation with the Washington State Department of Labor and Industries, conducted a survey of workers who recently received benefits. Nine percent (9%) stated that their employers "became upset or didn't believe them.".

As mentioned above, the total costs of workers' compensation and occupational injury and illness for the nation have elicited some interest. Total costs have also been estimated in Norway (20), Lebanon (21), and New Zealand (22). Specific diseases and injuries have also generated cost estimates: depression (23), needle-sticks (24), pneumoconiosis (25), non-fatal injuries (26), and homicide (27).

Workers' compensation data are useful in making the "business case" for investments in occupational safety and health. Lanoie et al. (28)

document increased profits to a firm that instituted an ergonomics program designed to reduce back injuries among warehouse workers. Tompa et al. (29) demonstrate cost-savings to instituting a different ergonomics program for workers in an auto parts manufacturing firm. A review of this "business case" literature is available (30).

Studies of workers' compensation markets might be useful for national debates about all medical costs (not just workers' compensation) and policies. For example, one recent workers' compensation study tried to identify reasons for the great increase in medical costs for the most severe injuries from 1996 to 2002 (31). A different study suggested that single-payer workers compensation carriers in Ohio and Washington, for example, might be more efficient than private insurance markets in the majority of other states (32). The national debate on costs also involves cost-shifting. For example, much discussion has focused on persons with insurance paying for persons without insurance. That is, some persons without insurance will nevertheless inevitably get sick and go to expensive emergency rooms for care. If the uninsured cannot pay out-of-pocket (as frequently happens), hospitals may pad the expenses of persons with insurance to cover the hospitals' loss in the emergency room. Different types of cost-shifting also apply to workers' compensation. In 2001, Medicare established the Workers' Compensation Medicare Set-Aside Arrangement (WCMSA) (33). This legal arrangement allows Medicare to recoup expenses that should have been paid by workers' compensation insurance carriers. (34). One study suggests the dollar amount carriers are being asked to set-aside is increasing in recent months (35). Another study suggests that the dollar amount for set-asides may need to grow a great deal to recoup Medicare expenses for occupational disease (36). Estimates suggest that workers' compensation is missing roughly 46,000 to 93,000 deaths and $8 billion to $23 billion in medical costs per year.

Future. There are numerous data-collection efforts and studies that can be imagined. Perhaps among the most cost-effective data-collection efforts would be to require that as many federal health services data sets as possible begin collecting workers' compensation data. These would include, for example, the Healthcare Cost and Utilization Project (HCUPnet) (37) and the Medical Expenditures Panel Survey (MEPS) (38). Both HCUPnet and MEPS are widely used and respected by health services researchers. Whereas they contain detailed information about insurance plans such as Veterans Administration, Medicare, Medicaid, CHAMPUS, among others, they do not have data on workers' compensation. It would be relatively easy to add a "box" for data-collectors to check if, for example, the hospital bill was paid by a workers' compensation carrier.

A Bureau of Labor Statistics (BLS) data source from the past ought to be resurrected: the Supplementary Data System (SDS). The SDS collected data from 7-20 state workers' compensation systems on Nature, Type and Source of injury as well as associated costs (39).

A promising, if difficult, area for future research would be to estimate the amount of fraud. I am unaware of any scientific study of fraud in workers' compensation. The U.S. General Accounting Office under the W. Bush administration estimated unemployment insurance (UI) fraud to be 2.16% of all UI spending (17). But any analysis of workers' compensation should also account for fraud by insurance companies. This amount is likely to be significant. These companies, after all, have the same profit motive as non-workers' compensation, HMOs, and private health insurance companies that middle-class Americans so frequently deride.

Stapleton and Thornton (40) have recently called for a National Disability Data System. This system would first pull-together exiting data from various sources on persons with disabilities. Second, it would begin to collect detailed, longitudinal data on Americans with disabilities. From our perspective, this proposed system should also collect data on whether and how the disability was job-related and whether workers' compensation was involved. What is amazing about the Stapleton and Thornton (40) "policy brief" are two things. First, no mention is ever made of workers' compensation. Second, in their Table 1 (which indicates possible databases), no workers' compensation or OSHA source is listed. For example, none of the following are listed: BLS Survey of Occupational Injuries and Illnesses (SOII); the National Council on Compensation Insurance (NCCI) data sources; state workers' compensation reports; the National Academy of Social Insurance reports; or the BLS Supplementary Data System.

Here is a list of possible future studies:

1. It would be useful to expand existing research on specific injuries and illnesses (20-27). These topics would allow workers' compensation studies to achieve greater recognition in the medical, public health, and epidemiological literature. That literature would reflect costs and benefits of interventions for carpal tunnel syndrome, hepatitis B and C, dermatitis, hearing loss, back injuries, motor vehicle injuries, homicides and assaults, and needle sticks, among many others. Given the rich data that are available from every state, from BLS, and, in time, from the large national data sets, we are well-positioned to develop the "gold standards" for economic studies involving these diseases and injuries that are of great interest to research-physicians, epidemiologists, and others.

2. A future study might try to document the extent to which hip and knee injuries during a person's working life (and paid by workers' compensation insurance) can be risk factors for hip or knee replacements during retirement (and paid by taxpayers via Medicare).

3. Given the rapid expansion of contingent workers, greater research needs to address whether and how much workers' compensation is covering this group.

4. The workforce is becoming more obese; yet fewer workers are smoking. Do these trends have offsetting medical cost effects?

5. What accounts for the long-term secular increases in workers' compensation costs? Number of injuries is on a secular decline. Even number of injuries resulting in more than 31 days of work-loss is on a secular decline. Are insurance companies heavily invested in stock markets and junk bonds? Are stock, bond, and real estate losses responsible workers' compensation insurance carriers' requests for higher rates?

6. To what extent does the BLS SOII undercount non-fatal injuries and illnesses? To what extent does the SOII include or exclude workers' compensation cases? Perhaps pilot projects should be sponsored that would include interviewing injured workers. Ultimately, a revision of SOII is in order and the new SOII should, in part, rely on input from workers or their associations, including unions.

7. Social Security records could be matched to past employment to determine if persons employed in certain industries for a specified minimum amount of time were more or less likely to develop cancer, circulatory disease, COPD, renal or neurological disease. Similar data was being assembled by NIOSH in the 1990s under the title "Mortality by Occupation, Industry, and Cause of Death: 24 Reporting States (1984-1988)" (41).

8. Finally, here is a suggestion for all future research: we need more transparency. Science does

not progress on faith. Science does not progress by researchers presenting results and saying, in effect, "trust me, I used reliable data and methods." Both data and methods must be available for other researchers to view.

Comments on Data Sources

I will comment on some data sources. Let me first offer caveats. These comments are personal, reflecting only my experience. The first caveat is that I have by no means contacted or dealt with most people, agencies, or firms that provide data, so many of them are omitted from these comments. The second caveat is that my dealings with people, agencies, and firms are limited. Over the years, personnel change and people become more or less receptive to inquiries from outside researchers.

There are many data sources. Some are excellent sources for published data, some for "special requests" that involve staff time, and some for both. Over many years of my experience, BLS is excellent for both. BLS, however, does not provide workers' compensation data, per se.

Virtually all states have agencies that collect some workers' compensation data. Most publish short reports. A North Carolina Industrial Commission website provides websites, addresses and phone numbers for these agencies (http://www.ic.nc.gov/ncic/pages/wcadmdir.htm, accessed August 20, 2009). In 2002 and 2003, I tried to contact every state agency to make a "special data request." Most were not able to help me. A number were, however. Personnel from the following states were very helpful: California, Florida, Wisconsin, Virginia, Washington, Minnesota, Maine, Colorado, Connecticut, Kansas, Mississippi, Missouri, New Mexico, North Carolina, and Oregon. The federal Office of Workers' Compensation also provides excellent published data and is generally helpful regarding special requests. (A list of specific contacts in each state is available from the author.)

Regarding private firms and institutes, I have benefited greatly from published data from the National Academy of Social Insurance, the Workers' Compensation Research Institute, the Upjohn Institute, Liberty Mutual Insurance Company, and the National Council on Compensation Insurance (NCCI). Data from the NCCI, however, are expensive. But even setting aside expenses, I have not been successful in acquiring data through "special requests" from the National Council on Compensation Insurance (NCCI). This is unfortunate since NCCI has arguably the best cost data available for most states.

One last point. Although the NCCI Annual Statistical Bulletin and NCCI State-of-the-Line reports have great amounts of data, I have never been able to find a national workers' compensation estimate in them.

Conclusion

Let me again reiterate the take-home messages. First, unfortunately, we do not get the respect we deserve. We should nevertheless keep in mind that we are advancing scientific knowledge that is critical for injured workers, their families, unions, employers, insurance carriers and governments. Workers' compensation should be an important part of the national debate on re-arranging the financing and delivery of medical care. Second, research fields are wide open. Many of the new questions will likely derive from cross-discipline collaboration. Working together, physicians, epidemiologists, public health researchers, and social scientists will likely produce more significant research than any one of these groups working alone. Third, numerous sources of data are available, but many opportunities exist for improvements in the quality of and access to these data. Moreover, existing highly regarded national datasets collected outside BLS and workers' compensation systems (e.g., MEPS and HCUP) could be expanded to include information on workers' compensation.

References

1. National Academy of Social Insurance. Workers Compensation Report for 2006. http://www.nasi.org/usr_doc/NASI_Workers_Comp_Report_2006.pdf Accessed July 27, 2009

2. Liberty Mutual. http://www.ergoweb.com/news/detail.cfm?id=413 Accessed July 27, 2009

3. Meropol NJ, Schulman KA. Cost of cancer care: Issues and implications. Jo.Clinical Oncology. 2007;25, 2: 180-186.

4. American Diabetes Association. Economic costs of diabetes in the US in 2007. Diabetes Care. 2008;31(3): 1-20.

5. Alzheimer's Association http://www.alz.org/national/documents_Report_2007FactsAndFigures.pdf

6. Budget of the US Government, Fiscal Year 2006 http://www.whitehouse.gov/omb/budget/fy2006/

7. Loeser JD, Henderlite SE, Conrad DA. Incentive effects of workers compensation benefits – A literature synthesis. Med Care Research Rev. 1995; 52(1):34-59.

8. Butler RJ, Worrall JD. Workers compensation – benefit and injury claims rates in the seventies. Rev Econ Statist. 1983;65(4):580-589.

9. Boden LI. Workers' compensation in the United States: High costs, low benefits. Ann Rev Public Health. 1995;16:189-218.

10. Boden LI, Reville RT, Biddle J. The adequacy of workers compensation cash benefits. In Workplace Injuries and Diseases, Prevention and Compensation: Essays in honor of Terry Thomason, edited by Roberts K, Burton JF, Bodah MM. Kalamazoo, MI: WE Upjohn Institute for Employment Research. 2005

11. Leigh JP, McCurdy SA. Differences in workers' compensation disability and impairment ratings under old and new California law. 2006 J Occup Environ Med 48(4): 419-425

12. Brigham CR. Permanent Impairment—Disability Rating Study: Impact of the AMA Guides to the Evaluation of Permanent Impairment, Fifth Edition on Permanent Disability Ratings in the State of California. Exhibit 2 in July 1, 2005, WCIRB Regulatory Filing as Amended May 19, 2005. Workers Compensation Insurance Rating Bureau

13. CHSWC, 2007, http://www.dir.ca.gov/chswc/Reports/memo_on_new_ratings_through_june_30_07_revised_aug_9.pdf Accessed July 27, 2009

14. Leigh JP, Miller TR. Ranking occupations based upon the costs of job-related injuries and diseases. J Occup Environ Med. 1997;39(12):1170-82.

15. Leigh JP, Waehrer G, Miller TR, Keenan C. Costs of occupational injury and illness across industries. Scand J Work Environ Health. 2004 ;30(3):199-205.

16. . Biddle J, Roberts K, Rosenman KD et al. What percentage of workers with work-related illnesses receive workers' compensation benefits? J Occup Environ Med. 1998;40(4):325-331.

17. Leigh JP. Expanding research on the economics of occupational health. Scand J Work Environ Health. 2006 Feb;32(1):1-4.

18. Lakdawalla D, Reville RT, Seabury S. How does health insurance affect workers' compensation filing? RAND Institute for Civil Justice Working Paper No. WR-205-1-ICJ. Santa Monica, CA. April 2005.

19 Gilmore Research Group. Survey of Injured Workers: Claim Suppression and Direction of Care. Prepared for State of Washington, Department of Labor and Industries. June 2007.

20. Rognstad K. Costs of occupational accidents and diseases in Norway. European J Operational Res. 1994;75(3):553-566.

21. Fayad R, Nuwayhid I, Tamin H, Kassak K, Khogali M. Cost of work-related injuries in insured workplaces in Lebanon. Bull World Health Organization. 2003;81(7):509-516.

22. 't Mannetje A, Pearce N. Quantitative estimates of work-related death, disease and injury in New Zealand. Scand J Work Environ Health. 2005;31(4):266-276.

23. Conti DJ, Burton WN. The economic impact of depression in a workplace. J Occup Environ Med. 1994;36(9):983-988.24.

24. Fisman DN, Mittleman MA, Sorock GS, Harris AD. Willingness to pay to avoid sharps-related injuries: A study in injured health care workers. Am J Infection Control. 2002;30(5):283-287.

25. Liang YX, Wong O, Fu H, Hu TX, Xue SZ. The economic burden of pneumoconiosis in China. J Occup Environ Med. 2003;60(6):383-384.

26. Al-Dawood KM. Direct impact of non-fatal occupational injuries. Saudi Medical J. 2000;21(10):938-941.

27. Hartley D, Biddle EA, Jenkins EL. Societal cost of workplace homicides in the United States, 1992-2001. Am J Ind Med. 2005;47(6):518-27.

28. Lanoie P, Tavenas S. Costs and benefits of preventing workplace accidents: The case of participatory ergonomics. Safety Science 1996;24(3): 181-196.

29. Tompa E, et al An economic evaluation of a participatory ergonomics process in an auto parts manufacturer. *Journal of Safety Research* 2009;40(1): 41-47

30. Tompa E, et al Practice and potential of economic evaluation of workplace-based interventions for occupational health and safety Journal of Occupational Rehabilitation 2006;16 (3): 367-392.

31. Shuford H, Restrepo T, Beaven N, Leigh JP Trends in Components of Medical Spending Within Workers Compensation: Results From 37 States Combined. J Occup Environ Medicine 2009;51 (2): 232-238.

32. Leigh JP, Bernstein J. Public and private workers compensation insurance. J Occup Environ Medicine 1997;39(2): 119-121.

33. . Meifert P. The workers compensation Medicare set-aside arrangement: Protecting Medicare's interests. Journal of Legal Nurse Consulting 2007;18(3): 11- 15.

34 Centers for Medicare and Medicaid Services http://www.cms.hhs.gov/workerscompagencyservices/04_wcsetaside.asp Accessed July 27, 2009

35. Swedlow A. http://www.workcompcentral.com/pdf/2009/misc/setasidereport071509.pdf. Accessed July 27, 2009

36. Leigh JP, Robbins JA Occupational disease and workers' compensation: coverage, costs, and consequences. Milbank Q. 2004;82(4):689-721.

37. Agency for Healthcare Research and Quality. http://www.hcupnet.ahrq.gov Accessed July 27, 2009

38. Agency for Healthcare Research and Quality http://www.meps.ahrq.gov/mepsweb/ Accessed July 27, 2009

39. Bureau of Labor Statistics, BLS Handbook of Methods, Bulletin 2414 (Bureau of Labor Statistics, 1992), chapter 14. US Government Printing Office. Washington, DC, 1992.

40. Stapleton D, Thornton C. Is it time to establish a national disability data system ? Center for Studying Disability Policy. Mathematica Policy Research, Inc. Number 09-03, May 2009.

41. Frank E, Biola H, Burnett CA. Mortality rates and causes among US physicians. Am J Prev Medicine 2000 19(3): 155-159

Differences among State Workers' Compensation Laws and Regulations

Keith Bateman, Vice President, Workers' Compensation Property Casualty Insurers Association of America

Researchers working with workers' compensation data for the first time, whatever the source, need to not only learn about the data being used but also the workers' compensation systems from which the data flow. Unless you understand these system aspects relevant to your research question, you may misinterpret what the data show and draw erroneous conclusions.

Having observed workers' compensation systems for more decades than I care to reveal, I am still learning new nuances about how systems differ from a statutory, regulatory, interpretational, and operational perspective. The richness of the permutations in state workers' compensation laws is truly amazing. Even if studying some aspect of a single system, this knowledge is important. When doing cross-state comparisons, it is critical. It is not only a case of dealing with apples and oranges, but even what appears to be an apple may turn out to be a "love apple" i.e. a tomato rather than an apple.

Although on a panel on "Aggregating Costs and Evaluating Trends", my role is to focus on discussing differences among state laws and regulations. This is timely because there currently are no legislative mega trends in workers' compensation. The trends that are multi-jurisdictional are not statutory-decreasing frequency, increasing medical costs, and changing politics of workers' compensation. Instead, states are focusing on making minor legislative adjustments.

Before discussing differences in state workers' compensation systems, two preliminary points need to be made. First, features of workers' compensation systems interrelate. A number of years ago my former trade association undertook a study of the characteristics of the permanent partial disability structures of the 51 jurisdictions. Dr. Peter Barth, Professor Emeritus of Economics, University of Connecticut, was a consultant to the project. Extensive interviews were conducted of claimants' attorneys, defense attorneys, and administrative law judges. Early in the interview process, Dr. Barth and I realized that one cannot understand the treatment of permanent partial disability without understanding how it interacted with the state's handling of temporary total disability, temporary partial, and permanent total benefits so we had to broaden our scope. Forces external to workers' compensation may drive trends rather than legislative change.

In the balance of the paper, I will address a number of the differences in system features including system features that affect incident rates and frequency of claims and the type of benefits paid.

Another presenter will be discussing getting the denominator right. My discussion will focus on differences in whom and what is covered.

A number of provisions affect who is covered. One is which employers are subject to the act. In Texas, workers' compensation is not mandated, and, in theory in New Jersey as well. In more than three-fourths of the states, workers' compensation laws apply to employers of one or more employees. Others exclude private employers with fewer than 3, 4, or 5 employees. In some of those, special provisions require construction employers to be covered if employing one or more. Most state laws apply to public employers, but not all. Some exclude certain public entities, and some exclude categories of public employees. The most common are firefighters and police officers, frequently covered by a separate disability/pension law. In private employment, the common exclusions are for farm employers, or more frequently farm laborers, domestics, and casual workers. Even among the states having these exclusions, states vary. Some have a blanket exclusion; others limit the application of the exclusion depending on certain characteristics of either the employer or employee. In the case of farmers and farm laborers, the test is usually tied to the farm payroll. For domestics, the exclusion usually is tied to the domestic's recent employment history. However, a few look to the employer's employing history. Some states bar all or certain classes of individuals (such as clergy) working for a nonprofit. While few states bar illegal aliens from coverage, some states disqualify them from receiving benefits if they lied on their employment application and others deny them benefits tied to income loss because their post-injury earnings loss was the result of their immigration status, not the injury.

Some states have addressed by statute whether particular classes of individuals should be considered employees. Most involve classes of individuals whose employment status was frequently disputed. Common provisions involve real estate brokers and salespersons paid on commission, owner/operators of trucks, newspaper distributors, and jockeys. Many states have special provisions dealing with partners, executive officers, and limited liability corporation members. States vary as to whether they are excluded, included, allowed to opt in or opt out, or allowed to be covered on a voluntary basis. Other special provisions are found for family members, work study program participants, volunteer workers, inmates working in the prison or outside, sole proprietors, and persons working for nonprofits for other than wages.

A minority of states allow individual workers to waive coverage prior to injury. Some limit the waiver to certain occupational diseases.

The injuries and diseases covered by law vary. Some cover all injuries and occupational diseases. Others cover only 'injuries by accident". Do not simply rely on the statutory language. Court decisions in states having such provisions have produced very differing interpretations.

Determining whether states cover mental injuries when no physical injury is involved requires looking at both statute and interpretation. Treatment of cumulative injuries also varies, and again it is important to look beyond the statute to its interpretation. The same is true for occupational diseases. Of note is that states may deal with occupational diseases either through their workers' compensation statutes or through a separate occupational disease act. When assessing a state's treatment of occupational disease, it is important to also review the statute of limitations applicable to disease, proof requirements, and benefits payable.

Also affecting the number of injuries eligible for cash benefits is the state waiting period before eligibility for cash benefits. States tend to cluster around either a 3- or 7-day waiting period, with a half dozen having a 5-day period. Most states provide benefits that are retroactive to the first day of disability after a certain time. The retroactive period varies from five days to six weeks with 14 days (or two weeks) being the most common followed by three weeks.

If relying on state agency data about injuries reported to it, you need to know that state reporting requirements differ. In some states, all injuries must be reported. In others, there must be lost time, ranging from one shift to ten days. In still others, reporting depends on whether medical treatment is provided. Some specify by whom or the type of treatment. One state requires, in addition to individual injury reporting, the reporting of all accidents when three or more employees are injured.

If your research involves the types of benefits provided by state laws, the differences discussed so far pale by comparison to the variations in state benefit provisions, particularly the unique workers' compensation concept of permanent partial disability (PPD). The most common claims are for medical only (60-80 percent of claims), which represent under 10 percent of the benefit dollars. The other types – temporary total (TTD), temporary partial (TPD), permanent partial (PPD), permanent total (PTD), and death benefits are where state benefit designs vary the most. Because of the paper's limited length, temporary partial disability and death benefits will not be discussed even though there are interesting variations in these benefit types.

Temporary total benefit provisions may base benefits on either gross wages or spendable earnings with the vast majority being two-thirds of gross wages. A few provide an additional amount tied to the number of dependents. Durations range from 104 weeks to the duration of total disability.

Sources of variation across all benefit types are maximums and minimums, and the computation of the injured worker's average weekly wage. Researchers also need to understand state provisions and practices regarding the ability of the parties to compromise on benefits.

Nowhere have states been more inventive than in the design of permanent partial benefits. Most, but not all, states provide for scheduled (specific) and unscheduled permanent partial benefits. Typically, scheduled body parts involve the extremities. However, some schedule all body parts while one state schedules only the back and others also schedule internal organs and the spine. Among body parts scheduled, there is no standardization regarding the value of a body part. Some value the loss of an arm more than a leg, while others are the exact opposite. Schedules may apply only to loss of body parts; others cover loss of use as well. In the latter case, some cover partial loss, while others do not. How partial loss of use is measured also varies. Some states require use of the AMA Guides to the Evaluation of Permanent Impairment which can lead to anomalies when the weeks scheduled for the loss of the body part is inconsistent with the valuation found in the AMA Guides. The schedule benefit may be the only recovery permitted whereas in other states additional compensation is possible. For amputations, some states provide a higher scheduled amount than if loss of use is involved. Scheduled benefits may be paid in a lump sum while others may require it to be paid periodically. In the majority, the injured employee's pre-injury wage is a factor in the benefit while other states use the state average weekly wage or a fixed amount of it. State treatment is not uniform when an injury involves both schedule and non-scheduled body parts. Two-thirds of the states provide a separate benefit provision for disfigurement but differ about the extent to which it must impact future earnings.

State ingenuity reached its zenith in benefit designs for unscheduled permanent partial disability. Even experts do not always agree on categories and which states to put in them. There are states that use impairment as a basis of compensation or as a proxy for disability, with or without statutory adjustment for things like age and education. Many use the AMA Guides to measure it. Even in states which supposedly base impairment solely on the AMA Guides, practices vary. Others base the benefit on loss of (future) wage earning capacity which usually combines a one time assessment of the degree of impairment with vocational opinion about the impact on future earning capacity. A number of states compensate for wage loss resulting from injury. There is disagreement about whether states that do not provide for a periodic comparison between pre and post injury earnings should be in this category. Other states provide different benefits, depending on whether or not the injured worker has returned to work.

Unscheduled PPD benefits are most often tied to the injured workers average weekly wage. The percentage and maximums and minimums used may differ from that used for TTD, durational limits vary, some have thresholds for eligibility, others provide supplemental benefits upon benefit exhaustion in certain circumstances, a few allow the claimant to select the benefit type, others have an aggregate cap on TTD and PPD either in weeks or dollar amounts. Some states use schedules of benefits in which duration or amount is tied to the percentage of impairment or disability either on a straight basis or on a tiered basis, providing more weeks of benefits for the higher degrees of impairment or disability. A few provide cost of living adjustments. This is only a partial list of the permutations.

In cases of permanent total disability (PTD), qualifying may be limited to those injuries specified by statute, may be a factual determination, or a combination of the two. Durations may be for life or until retirement or have a number of weeks limit. The status may be a one-time determination, subject to periodic review, or a review at a given point in time. Some states provide for cost-of-living adjustments that may or may not be capped.

One's mind can be numbed by this listing of differences. However, it is only differences that may affect the research question you are addressing that must be understood. The point being made is that simply looking at a data set without understanding of the context is a recipe for bad research.

National Averages of Employee Benefits and Employer Costs for Workers' Compensation

John F. Burton Jr., Rutgers University

This article identifies several sources of national data for employee benefits and employer costs for workers' compensation programs in the U.S. This article does not discuss data published by individual states, in part because often such data are not comparable across states, or multi-state data sets that do not encompass most states, such as the CompScope data published by the Workers Compensation Research Institute.

National Academy of Social Insurance: Benefits and Costs

The National Academy of Social Insurance (NASI) publishes an annual report providing information on benefits, coverage, and costs. The most recent edition (Sengupta, Reno, and Burton 2009) provides data for 2007. The NASI reports began in 1997 after the Social Security Administration discontinued publishing the data series, which contains annual estimates dating back to 1946.

Figure 1 provides an example of the NASI data. Benefits and costs per $100 of payroll increased during much of the 1980s, peaked in the early 1990s, and generally declined since then. Employers' costs declined from $2.17 of payroll in 1993 to $1.45 of payroll in 2007 – down a third – while benefits paid to workers dropped from $1.65 per $100 of payroll in 1992 to $0.95 of payroll in 2007 – down more than 40 percent.

The NASI reports contain national information on the number of workers and the amount of payroll covered by workers' compensation; the share of benefits provided by private carriers, state funds, and self-insuring employers; the proportion of benefits paid under large deductible insurance plans; the share of benefits accounted for by cash and medical benefits, and employers' costs. With the exception of employers' costs, these data are available for each state.

The NASI report is based on data from the insurance industry and from state agencies. However, only 28 of the 51 jurisdictions that allow employers to self-insure were able to provide data on benefits paid by self-insurers and only 7 of the 46 jurisdictions that allow carriers to write deductible policies were able to provide the amount of benefits paid under these policies. NASI used several methods to estimate the missing information. The lack of comprehensive data on self-insuring employers is one reason why NASI does not publish employers' cost data by state.

Bureau of Labor Statistics: Costs

The Bureau of Labor Statistics (BLS) publishes Employer Costs for Employee Compensation, which contains information on wages and salaries and benefits other than pay provided by employers, including workers' compensation. Information on private sector employees is available since 1986 and data for state and local government employees and for all non-federal employees are available since 1991. Since 2002, the BLS data are available on a quarterly basis. The most recent data used for this article (U.S. Department of Labor, 2009) was based on a sample of 13,600 establishments in private industry and 1,900 establishments in state and local governments. The BLS methodology and the procedure used to calculate workers' compensation costs per $100 of payroll are discussed in Burton (2008a: Appendix A).

Figure 2 presents the national BLS data on employers' costs for the private sector and for all non-federal employees as well as the NASI data on employers' cost for all employees. Except for 1986, the costs are higher in the BLS data than in the NASI data.

Figure 1. Workers' Compensation Benefits and Costs per $100 of Covered Wages, 1980-2007, NASI Data

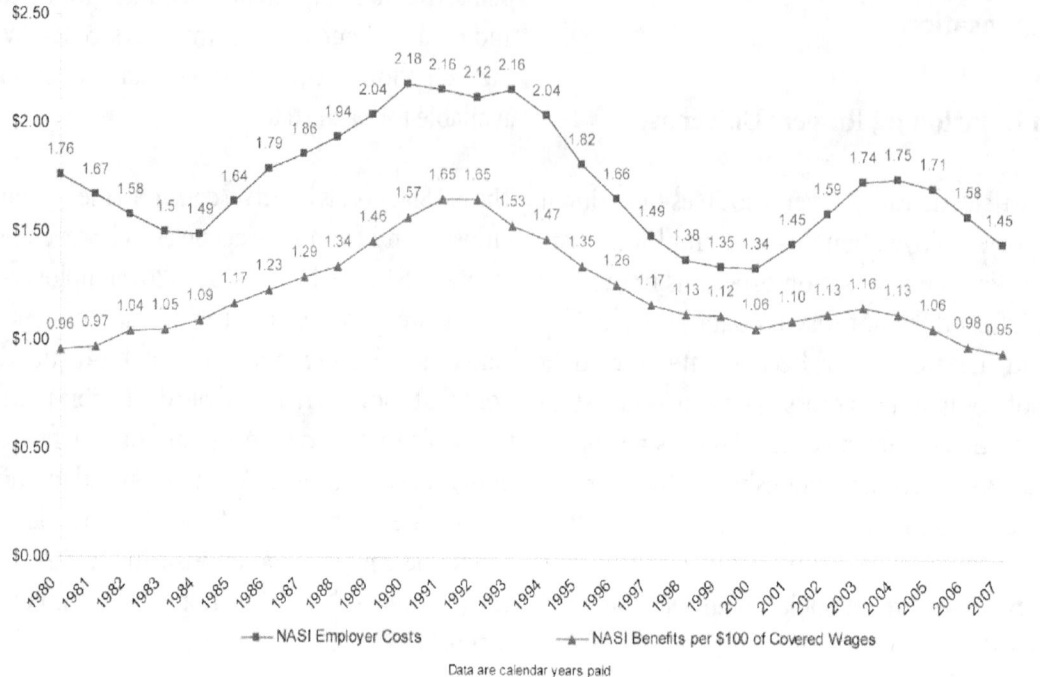

Figure 2.

Workers' Compensation Costs per $100 or Payroll, 1980 - 2008, NASI and BLS Data

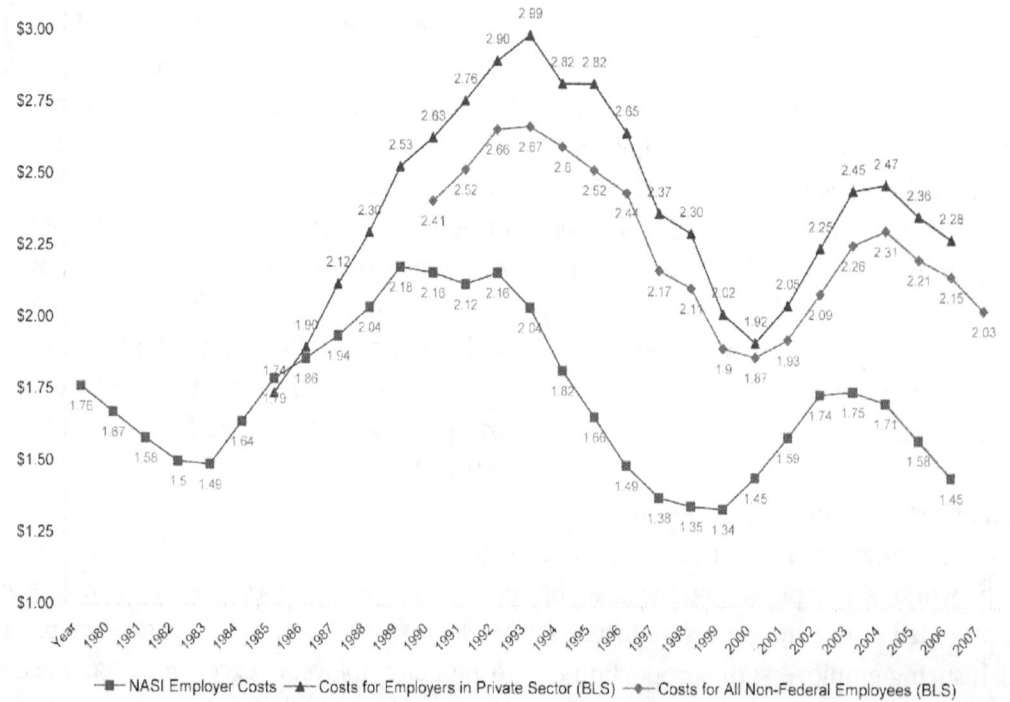

Burton (2008: 36-37) discusses the differences in the peak and trough years for the two data series.

The BLS data on employer costs in the private sector are available by industry, occupational group, establishment size, bargaining status, and for four census regions and for nine census divisions (Blum and Burton 2008a), but not for states.

Paid Benefits and Incurred Benefits

The NASI data on benefits are calendar year paid benefits, which are the benefits paid to workers in a given year, regardless of when the injury or illness occurs. The data included in the NASI reports on covered workers and total (cash plus medical) benefits can be used to calculate paid benefits per 100,000 workers for the nation (shown in Figure 3) and for individual states.

The National Council on Compensation Insurance (NCCI 2008) annually publishes data on frequency of claims for four types of cash benefits and the averages of medical and cash benefits for those claims, plus the frequency and average benefits for claims providing only medical benefits. The benefits are incurred benefits, which are the benefits for injuries that occurred in a specified policy period paid prior to a specified report date plus liabilities for future benefits for those injuries. The Workers' Compensation Policy Review (WCPR) used the NCCI data from the first reports of the injuries to construct national and state data on frequency, average benefits per claim, and benefits per 100,000 workers on a calendar year incurred basis.

The national averages of NCCI/WCPR incurred benefits per 100,000 workers for 1984-2007 (Burton and Blum 2008) are also shown in Figure 3. The NCCI/WCPR incurred benefits are consistently greater than the NASI paid benefits, which may be due to the inclusion of data for self-insuring employers in the NASI data and the limitation of the NCCI/WCPR data to employers who purchase insurance from private carriers or from some of the competitive state funds.

Figure 3.

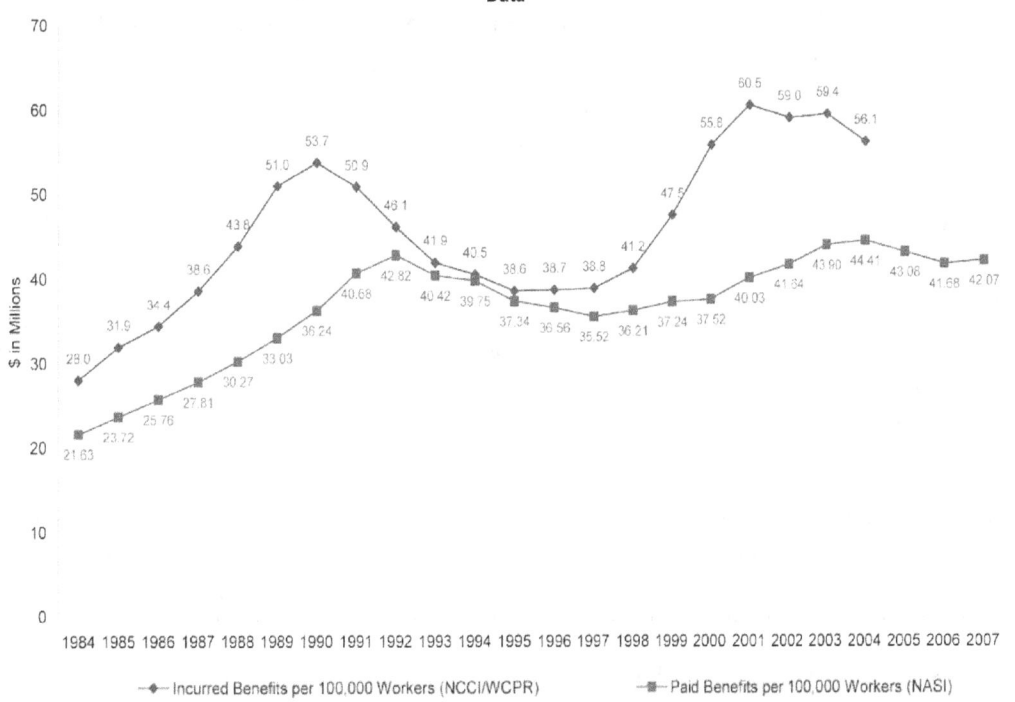

Figure 4.

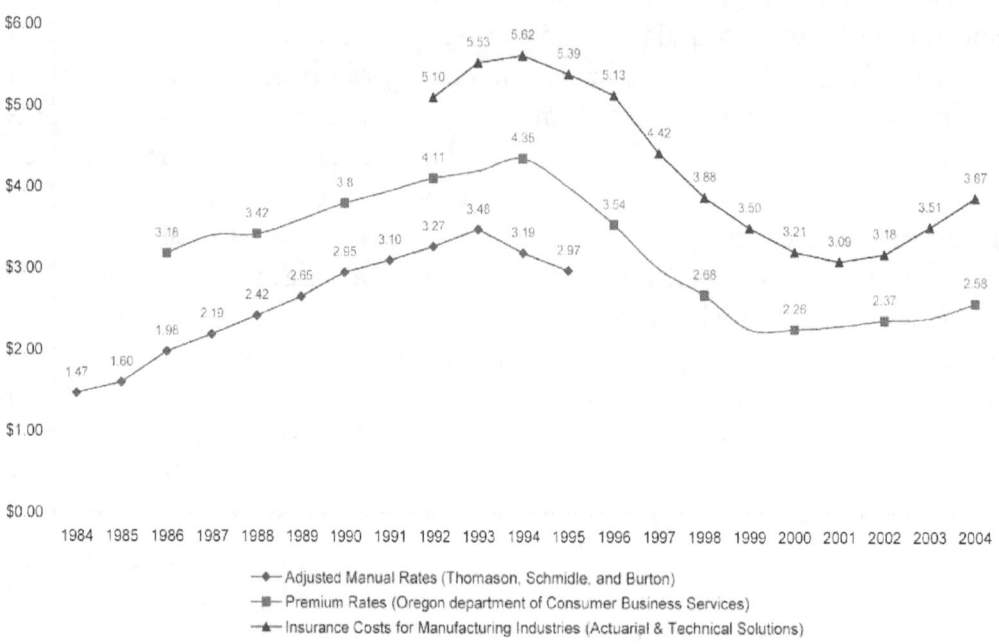

Figure 4
Workers' Compensation Insurance Premiums per $100 of Covered Wages, 1984-2004, Three Variants

The national NASI data can be disaggregated by cash or medical benefits and by type of insurance arrangement (private carriers, self-insurance, or state fund) since 1980. The state NASI data can be disaggregated by type of insurance arrangements since 1980 and by cash or medical benefits since 1987.

The national and state BLS/WCPR data can be disaggregated by cash or medical benefits, by type of cash benefits, and by medical-only benefits since 1985. State data are missing for states without private carriers and for some years for states with private carriers. Since 1985, the number of jurisdictions (including the District of Columbia) with data varied between 44 and 47 (Burton and Blum 2008: Table 2).

The latest NASI report (Sengupta, Reno, and Burton 2009: 25-27) contains caveats on using the NASI paid benefits to compare benefit adequacy across states and on using benefits data to compare employer costs across states, and most of the caveats (or admonitions) are also applicable to using the NCCI/WCPR incurred benefits for these purposes. A major limitation of both sets of data is that the data do not control for interstate differences in industry or occupational composition.

Insurance Premiums as Measures of Employers' Costs

Three studies have measured the employers' costs of workers' compensation by calculating the insurance premiums per $100 of covered wages. Each of the studies produced national averages, as shown in Figure 4, as well as data for most or all states and each used the same set of insurance classes and the same distribution of payroll among the classes in order to control for interstate differences in industry mix.

Thomason, Schmidle, and Burton (2001) calculated annual adjusted manual rates per $100 of payroll for 71 insurance classes for 1975 to 1995. The methodology, explained in excruciating detail in Appendix C of their study, began with manual rates (or pure premiums), and adjusted for factors such as expense constants, experience rating, deviations, schedule rating, dividends, and the different rates in voluntary and assigned risk markets, to arrive at adjusted manual rates. The number of states varied between 42 and 48, depending on the year. The data are reproduced in Burton and Blum (2005: Table 7).

The Oregon Department of Consumer & Business Services (Oregon) has calculated average premium rates per $100 of payroll for 50 insurance classes in alternate years since 1986. The methodology, described by Burton and Blum (2005: 35-37), begins with manual rates or pure premium, and adjusts for different expense loading factors in voluntary and residual markets. However, the premiums were not adjusted for factors such as expense constants, experience rating, deviations, schedule rating, and dividends, which usually reduce the premiums paid by employers. The 1986 to 2004 data for all 50 states are reproduced in Burton and Blum (2005: Table 6).

Actuarial & Technical Solutions, Inc. (A&TS) has calculated annual average insurance costs per $100 of payroll for manufacturing industries since 1992. The methodology, described by Burton and Blum (2005: 29-30), begins with manual rates and adjusts for expense constants, experience rating, deviations, schedule rating, and dividends, to produce the insurance costs. The 1992 to 2005 data for the 46 states with private carriers as of 1992 are reproduced in Burton and Blum (2005: Table 5)

The results in the overlapping years for the three measures of workers' compensation insurance premiums indicate that the adjusted manual rates from Thomason, Schmidle, and Burton are lower than the Oregon premium rates, which may be explained by the absence of some adjustment factors in the Oregon methodology, such as deviations and dividends. The A&TS insurance costs are higher than the two other measures of insurance premiums, which may be explained by the use of manufacturing industries only, which tend to have relatively high insurance rates.

The Links Between Workers' Compensation and Workplace Safety

There are numerous ways that the links between workers' compensation and workplace safety can be examined. Here are two examples.

Improved workplace safety can result in reduced workers' compensation benefits for workers and lower costs for employers. However, Guo and Burton (2010) examined the NCCI/WCPR state data on incurred cash benefits (similar to the national data on total incurred benefits shown in Figure 3) and found that the decline in cash benefits during the 1990s was due more to tightening eligibility rules for workers' compensation benefits than to reductions in the injury rate.

The relationship between the workers' compensation insurance arrangements used by states and workplace safety was examined by Thomason, Schmidle, and Burton (2001) (using state data corresponding to the national data in Figure 4) and produced paradoxical results. They found that injury rates were higher in states with exclusive state funds compared to states with private carriers, but that injury rates were lower in jurisdiction with competitive state funds than in states with only private workers' compensation carriers. This paradox surely warrants substantial resources for researchers to resolve.[2]

References

Blum, Florence and John F. Burton, Jr. 2008a. . "Workers' Compensation Costs in 2007: Regional, Industrial, and Other Variations." Workers' Compensation Policy Review, Vol. 8, Issue 2 (March/April) 3-14.

Blum, Florence and John F. Burton, Jr. 2008b. "Workers' Compensation Benefits: Frequencies and Amounts in 2004." Workers' Compensation Policy Review, Vol. 8, Issue 4 (July/August): 3-27.

Burton, John F., Jr. 2008. "Workers' Compensation Costs for Employers 1986 to 2007." Workers' Compensation Policy Review, Vol. 8, Issue 1 (January/February): 19-39.

Burton, John F. Burton, Jr. and Florence Blum. 2005. Workers' Compensation Compendium 2005-06: Volume Two. Princeton, New Jersey: Workers' Disability Income Systems, Inc.

Burton, John F. Burton, Jr. and Florence Blum. 2008. "Workers' Compensation Incurred Benefits: 1985-2004." Workers' Compensation Policy Review, Vol. 8, Issue 5 (September/October 2008): 3-23

Guo, Xuguang (Steve) and John F. Burton, Jr. 2010. "Workers' Compensation: Recent Developments in Moral Hazard and Benefit Payments." Industrial and Labor Relations Review, Vol. 63, No.2

National Council on Compensation Insurance. 2008. Annual Statistical Bulletin: 2008 Edition. Boca Raton, FL: National Council on Compensation Insurance.

Thomason, Terry, Timothy P. Schmidle, and John F. Burton, Jr. 2001. Workers Compensation: Benefits, Costs, and Safety under Alternative Insurance Arrangements. Kalamazoo, MI: W.E. Upjohn Institute for Employment Research.

U.S. Department of Labor, Bureau of Labor Statistics. 2009. Employer Costs for Employee Compensation – December 2008. USDL: 09-0247. March 12, 2009. Washington, DC: U.S. Department of Labor.

The articles from the Workers' Compensation Policy Review can be downloaded without charge from www.workerscompresources.com.

---

[2] The results from Thomason, Schmidle, and Burton (2001) are particularly paradoxical because they found that, after controlling for factors that influence employers' costs of workers' compensation insurance, there were no statistically significant differences in the costs of workers' compensation insurance between exclusive-state-fund jurisdictions and states that allow private carriers, but that states with competitive state funds have workers' compensation insurance rates that are considerably higher – nearly 18 percent – than the insurance rates in states with only private carriers.

Learning from Workers' Compensation Claims Triangles

Frank A. Schmid, National Council on Compensation Insurance, Inc.

Workers' compensation is a long-tailed line of insurance, where claims generate a stream of payments that may continue on for several decades. Aggregate claims triangles record for each accident year the development of the ensuing annual (incremental) payments on the set of claims recorded in this accident year. These payments are typically broken down into their indemnity and medical components. The development pattern of incremental payments offers information on the consumption path of medical and indemnity services as the set of claims recorded in a given accident year matures. Further, by separating changes in consumption from changes in price, the impact of inflation may be discerned—inflation may be legally stipulated (e.g., cost-of-living adjustments of indemnity payments), the result of legislative action (e.g., changes to medical fee schedules), or originate in general inflation.

Exhibit 1 displays a stylized claims triangle, where {i,j} indicates a payment made in development year j for the set of claims recorded in accident year i. In calendar time, the first development year always equals the accident year. The latest recorded payments (which, in Exhibit 1, are the payments for calendar year 2008) are printed gray.

Exhibit 2 shows that there are three time dimensions to a triangle: Volume may change with the accident year, consumption of (indemnity or medical) services may change with the time distance to the accident, and there may be effects associated with the calendar year (among which are legislative reforms that affect claims retroactively, and inflation).

Because claims may generate payments for many decades, studying the payment pattern until claims closure requires triangles that comprise many decades of accident years—only very few such triangles are available for research purposes. There is an SCF (State Compensation Fund) Arizona indemnity triangle, which comprises 74 years of development, the accident years range from 1930 through 2003. Further, there is an SAIF (State Accident Insurance Fund, Oregon) triangle, which consists of the medical component of permanent disability claims; the accident years run from 1926 through 2005. Schmid (2009) analyzes these triangles using a Bayesian statistical model. Due to a dearth of data for accident years 1930 1937, these eight years are excluded from the

Exhibit 1: Stylized Aggregate Claims Triangle of Development Year by Accident Year

	Dev. Year 1	Dev. Year 2	Dev. Year 3	Dev. Year 4
Accident Year 2005	1,1	1,2	1,3	1,4
Accident Year 2006	2,1	2,2	2,3	
Accident Year 2007	3,1	3,2		
Accident Year 2008	4,1			

Exhibit 2: Claims Triangle Architecture

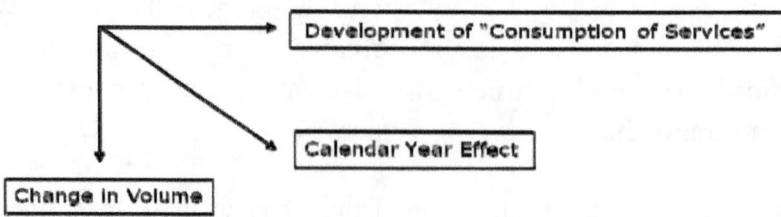

Exhibit 3: Rate of Decay in Modern Indemnity Services

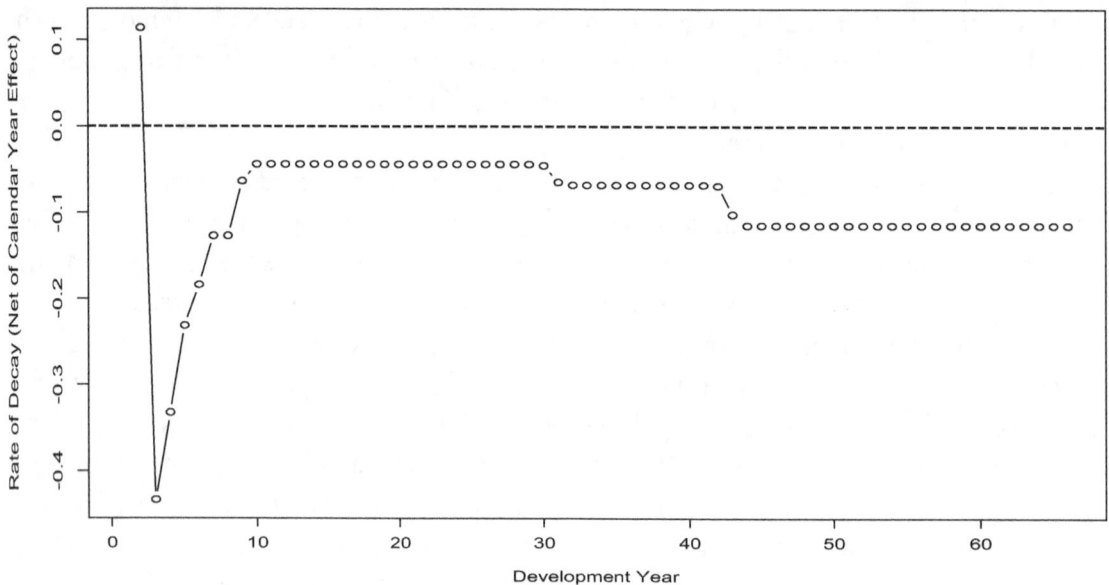

Exhibit 4: Probability of Observing an Indemnity Payment

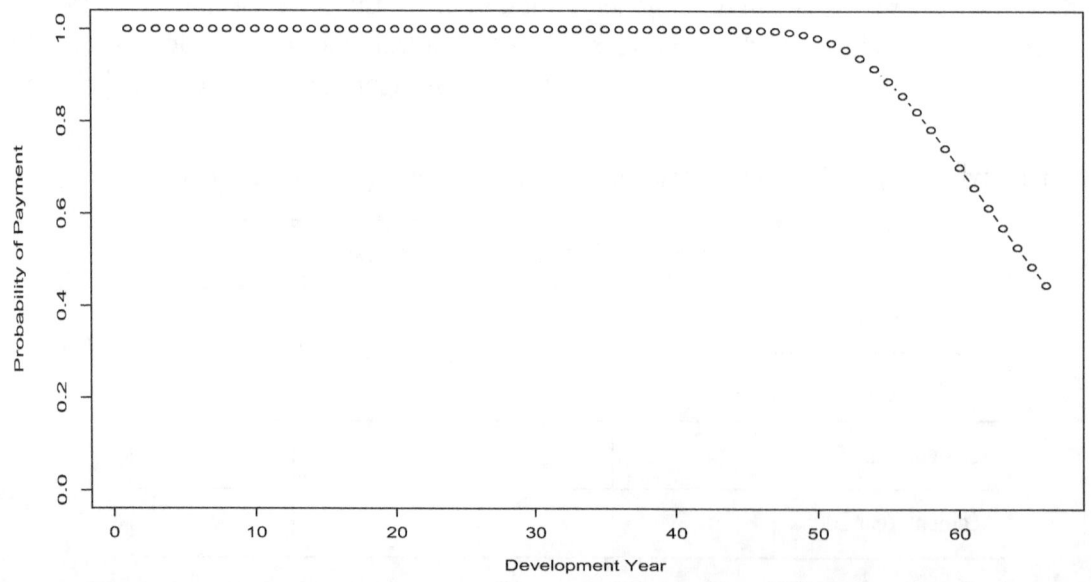

Exhibit 5: Rate of Decay in Consumption of Medical Services

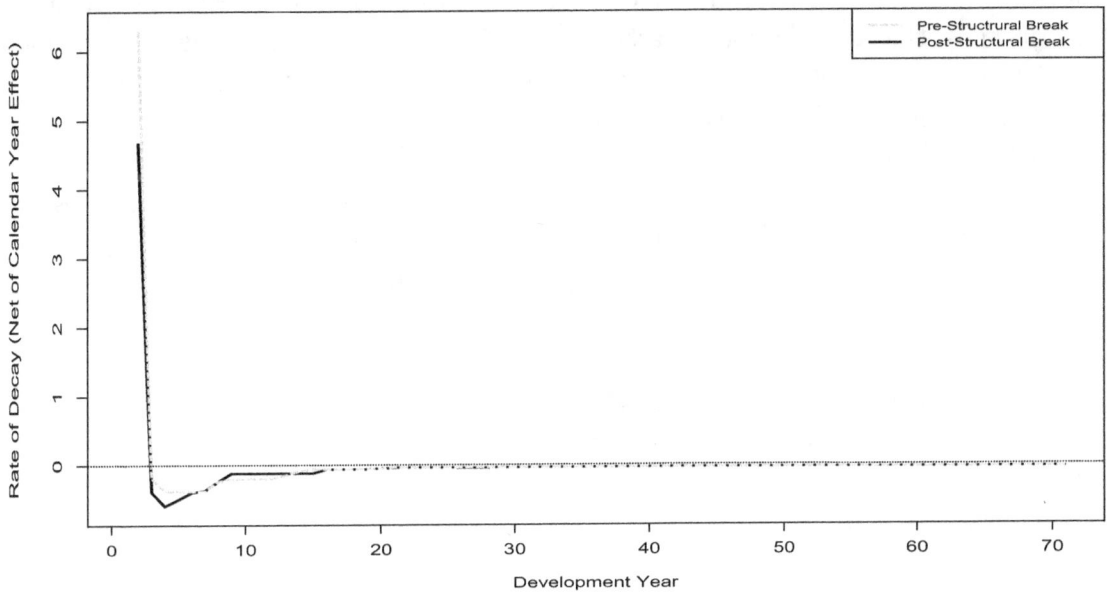

Exhibit 6: Probability of Observing Medical Payment

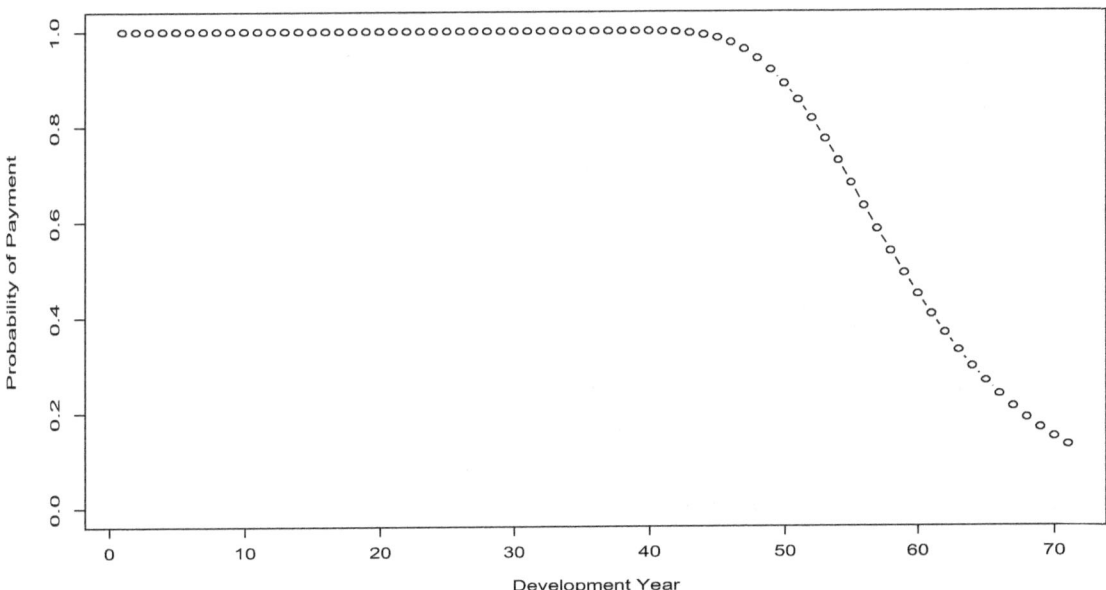

analysis of the SCF triangle, thus reducing the data to 66 development years. Similarly, due to the sparseness of the data, the first nine accident years (1926 1934) of the SCF indemnity triangle are discarded, thus reducing the set of observations to 71 development years.

Exhibit 3 displays the rate of change in the consumption of indemnity benefits. Consumption is defined as payments, adjusted for calendar year effects (which, where applicable, include cost of living adjustments). The exhibit shows that the decline of consumption quickens as claims develop, following the rate of mortality of the cohort of injured claimants.

Exhibit 4 presents the probability of observing a nonzero indemnity payment in a given development year. The trajectory of this probability was treated as uniform across accident years. To the degree that longevity has improved over the past couple of decades, this trajectory has to be shifted to the right when simulating the lifetime consumption for cohorts of claimants of more recent accident years.

Exhibit 5 shows the rate of change in the consumption of medical benefits. Consumption is defined as incremental medical payments adjusted for calendar year effects (which include inflation). Most interestingly, the rate of decline of consumption stabilizes around development year 20, which implies that from then on, the increase in the rate of mortality is partially (and at constant proportion) offset by an increase in the consumption of medical services among the remaining claimants. Note that the cost containment reform of 1990 has led to an accelerated run-off during the first couple of development years.

Exhibit 6 displays the probability of observing a nonzero medical payment in a given development year. As discussed, to the degree that longevity has improved over the past couple of decades, this trajectory has to be shifted to the right when simulating the lifetime consumption for cohorts of claimants of more recent accident years.

Reference
Schmid, Frank A. (2009) The Workers Compensation Tail Revisited, First Draft: January 2009; Revised: September 2009, http://www.ncci.com.

Identifying and Tracking Trends in Workplace Injuries and Illnesses – Opportunities and Challenges in Using Workers' Compensation Rating Bureau Data

Harry Shuford, Practice Leader and Chief Economist National Council on Compensation Insurance, Inc.

Background

It is correctly perceived that workers' compensation (WC) rating bureaus have a lot of data. There is a likely perception that these data could be of value in identifying and tracking trends in workplace injuries and illnesses. This paper offers some insights into the opportunities and challenges in working with the data typically collected and analyzed by WC rating bureaus such as NCCI.

The Role of a WC Statistical Agent/Rating Bureau in the WC System

A WC insurance rating bureau is an organization licensed by an insurance regulator or created by state statute that serves multiple roles. As a statistical agent the entity is responsible for collecting and organizing financial and claims data required by the insurance regulator to support its regulatory responsibilities. As a rating/advisory organization the entity analyzes the data and submits reports – typically in the form of annual rate or loss cost filings – to the insurance regulator; these contain actuarial analysis and recommendations for the coming year's premium rates. In most states insurance law typically requires that premium rates should not be excessive, nor inadequate, nor unfairly discriminatory. These criteria are consistent with the Casualty Actuarial Society's Statement of Principles Regarding Ratemaking.

Aggregate ratemaking estimates the adequacy of current rates ("experience") and projects anticipated trends in the frequency of claims (comparable to BLS injury rates) and the severity of claims (the average cost – medical and indemnity separately per claim.) The objective is to achieve an average rate level that is neither excessive nor inadequate overall.

Class ratemaking allocates premium costs according to relative class and group experience (i.e. rates for contracting are higher than average due to both higher frequency and higher severity for the typical firm in that industry group.) The allocation based on class and group experience is also intended to help achieve rates that are neither excessive nor inadequate with the additional objective of not being unfairly discriminatory.

Experience rating calculates debits or credits (termed "experience mods" or modifications) for individual policyholders relative to other policyholders in their industry group. This experience rating is intended to support rates that at the individual policyholder level are not excessive, nor inadequate, nor unfairly discriminatory. (Because of the potential year-to-year volatility due to limited statistical credibility of individual policyholder experience smaller policyholders typically do not qualify for experience rating.)

Ratemaking in the form of "legislative pricing" estimates the likely impact on claims costs and therefore premium rates resulting from changes in benefit levels, medical fee schedules, and other system changes. If the changes are enacted the estimated impact will be included in a rate/loss cost filing (either as part of the annual aggregate filing or as a stand-alone "law only" filing.) Again the objective is to maintain rate/loss costs that are neither inadequate nor excessive.

Research conducted by a rating bureau generally is designed to support ratemaking by providing insight into factors that contribute to, for example, poor experience and/or loss cost trends.

The Types of Data Typically Collected by a WC Rating Bureau

Aggregate ratemaking for WC relies on financial data - total number of lost time claims, total medical and indemnity payments, total medical and indemnity incurred costs (payments plus outstanding case reserves on open claims) – submitted annually by all WC insurers in a state. Currently the financial data submitted to NCCI covers the most recent 20 years – each year is reported separately and an update of the financial data of each year is reported at annual increments. For example, in 2009 there was one report for the year 2008 but there were 20 report periods for the year 1989. Typically financial data are provided for both "policy years," which track results for all insurance policies written in a calendar year (a policy year, therefore, covers 24 months as policies written in late December of one year cover claims that might occur in the following calendar year), and for "accident years," which track results for claims due to injuries or illnesses occurring in a calendar year.

Class ratemaking relies on analyzing the claims experience for each policyholder. The National Council on Compensation Insurance, Inc. (NCCI) receives these data – "unit statistical records" - from insurance carriers annually. There is a report for each policyholder for the year indicating the status of each claim associated with that policy: claims cost data – e.g. payments, incurred costs (i.e. payments plus case reserves); injury detail – e.g. part of body, nature and cause of injury; occupation of injured worker. A unit record for a given WC insurance policy is submitted annually for 5 years; this time frame recently was expanded to 10 years.

Experience rating calculates debits or credits ("experience mods") for individual policyholders based on their claims experience relative to other policyholders in their industry group. This analysis is based on the three most recent years of unit statistical records. This analysis of individual policyholder experience is more heavily influenced by frequency as the cost of individual claims is capped. This addresses some statistical concerns but also is thought to encourage increased attention on workplace safety.

Legislative pricing typically requires more detail than is available in unit records. A primary source of this detail is the "Detailed Claims Information" (DCI) data submission. This is a stratified random sample of the carriers' lost time claims. Compared with the unit records the greater DCI detail includes demographic information on the injured worker (date of birth, date of employment, date of injury, gender, nature of employment) as well as more detail on the specifics of the claim – indemnity: average weekly wage, weekly benefit, vocational rehab, closure date, return to work status; medical: hospital payments, other medical, degree of impairment. This is currently supplemented with a separate database with transactions detail – indemnity: checks including amount and time period covered; medical: payments including diagnosis, amount paid, service rendered, date of payment[1].

Limitations on the Use and Distribution of Data NCCI Collects

Data usage and other agreements place limits on the use of the data submitted by insurance companies to NCCI. For example, the data typically may not be distributed or shared with other parties except for limited, specified purposes and/or in an aggregated format.

Practical Challenges in Using WC Rating Bureau Data

There are also practical challenges to using the data described above. For example, these data are not particularly useful in trying to track occupational diseases and illnesses; due to the long span between exposure and manifestation few of these are reflected in current periods' data.

In terms of early warning of emerging illnesses and exposures there is a comparable problem – essentially trying to separate a signal (the emerging trend) from a remarkably diverse range of workplace injuries. A recent study of patterns related to carpal tunnel injuries determined that at the detailed diagnosis code level such injuries were the most common cause of lost time claims. However, even at number one they accounted for only 4% of all lost time injuries. A range of four back and two neck injury diagnoses dominated the top 10. The 20th ranking diagnosis (open wound of finger) accounted for barely 1%; there are another 330 diagnoses with at least one claim.

Other data sets also become sparse after separating data according to as few as two or three characteristics, making detailed analysis a challenge.

The WC experience of self-insureds businesses (typically larger employers) and self-insured government entities typically is not captured in the data submitted to rating bureaus.

Projects that Used WC Rating Bureau Data to Assess Trends and Patterns in Workplace Injuries

In spite of these limitations it is still possible to conduct meaningful research using bureau claims data. The focus primarily is on identifying and quantifying potential drivers of claims costs. The insights from these research findings are of value to underwriting, claims management, and safety/loss control.

NCCI is establishing a new "medical data call" that will include medical transactions detail on virtually all WC claims in NCCI states. There also will be an enhanced DCI data call. These data sets are intended primarily to support NCCI's analysis of proposed legislative changes that will impact WC claims costs and actuarially sound premium rates. They will be subject to the types of legal restrictions discussed in the next section.

Aging Workforce/Workers 65 and Older – It is generally accepted that older workers experience higher costs claims. In one study "slips, trips, and falls" was identified as a leading cause of workplace injuries for workers 65 and older, resulting in high levels of fractures and head injuries. In another, it was noted that the combination of relatively high frequency and relatively high severity results in the 35 to 44 age cohort having the highest average claim cost per worker.

Medical Utilization – From 1996/97 to 2001/2002 the medical cost of an average WC claim increased from just over $3,600 to more than $6,350; an increase over five years of more than 70%. Medical prices increased by 20% over the same period. An observer might take this as a sign that workplace safety had seriously deteriorated in the dot-com boom. An analysis of medical transactions data indicated that, indeed, the increases in prices paid for WC medical services tracked the growth in the medical CPI. There also was a modest increase in the share of claims with high cost diagnoses, due in part to a small increase in the share of older workers. At least half of the 70% increase, however, reflected the 35% increase in the number of billed medical treatments – for the same mix of diagnoses. The cause of this increase in utilization is unclear, but the dramatic growth in medical claims costs did not reflect a marked deterioration in workplace injuries.

Studies analyzing workplace violence, trends in the long term care industry, and traffic accidents are also available on ncci.com.

Anticipated Actions to Extract More Value from WC Bureau Claims Data

Safe Lifting Practices at Long Term Care Facilities – Academic researchers were interested in determining the impact on WC claims costs when long term care facilities implemented safe lifting programs and practices. Because this proposed research was closely related to evaluating potential changes in loss cost trends NCCI partnered with academic researchers in a joint project. The academic team developed data on the safe lift programs for individual institutions; NCCI matched the institutions with policyholder data and estimated the relationship between the "treatment" (i.e. safe lift programs) and claims experience. Preliminary results indicate that on average both frequency and total claims costs were lower at institutions with more robust safe lift programs. In addition it appears that for-profit and government owned institutions on average had better outcomes.

Obesity – This study was hampered by data limitations. In particular, the claims data did not include information on such factors as BMI. Instead, the study used a matched pair sample approach of claims that differed only in that one had a secondary or tertiary diagnosis of obesity whereas the other did not. Matching criteria included primary diagnosis, gender, industry, state, accident year, and age. Comparing cumulative payments 12 months after the date of injury indicated that the total medical costs for the claims where obesity was reported as a complicating factor were more than three times the total medical costs on the matched claims without such a diagnosis. It was equally revealing that the average of the ratios of the individual pairs was over thirty at 12 months; the interpretation – typically there are a large number of primary diagnosis injuries that are relatively low cost for non-obese patients but which are dramatically higher for otherwise comparable patients where obesity is a secondary/tertiary diagnosis. Another key finding: the differences in medical costs of these matched pairs of claims on average were 15% to 30% lower in states with mandatory utilization or mandatory bill review.

Job Flows – Frequency change is the key driver of changes in WC annual loss costs. The conventional wisdom in WC circles is that frequency increases in recessions as laid off workers choose WC benefits over unemployment. Economic analysis, however, clearly shows that the rate of change in frequency drops in recessions. There is also considerable concern in the WC industry that the 20 year decline in frequency cannot continue indefinitely; identifying the turning point will be critical to avoiding inadequate premium rates.

Frank Schmid recently completed a study that used Monte Carlo Markov Chain Bayesian modeling to provide dramatic insights into the dynamic relationship between changes in labor market conditions and changes in frequency as measured by BLS injury rates. In particular, he linked changes in job flows (termed "job creation" and "job destruction") to changes in injury rates. It appears, for example, that in a recession increases in layoffs (job destruction) are linked to upward pressure on frequency, just as industry members believe. However, this force is dominated by the much more pronounced downward pressure on frequency due to the coincident drop in the rate of new hires (job creation), arguably reflecting the observed higher injury rates of inexperienced workers (in recessions there are fewer of them.) Schmid's analysis also found that there was essentially no difference in the effects of layoffs at existing establishments and layoffs due to closings. In contrast, unlike new hires at existing establishments, new hires at new establishments actually seem to be linked to downward pressure on frequency growth. New technology, on average, appears to be safer technology. It seems reasonable to speculate that the ongoing replacement of old technology by new methods is a key factor in the long term decline in injury rates and claims frequency.

The studies of jobs flows and of obesity provide examples of how advanced statistical analysis can extract meaningful insights from sparse or otherwise limited data. Data mining tools such as WEKA offer enhanced opportunities to identify small but systematic changes over time – potentially isolating the signal embedded in noisy data.

Closing Thoughts

The data currently collected by WC rating bureaus is designed to support actuarial analysis and the promulgation of actuarially sound premium rates; rates that are "not excessive, not inadequate, and not unfairly discriminatory." The data have proved valuable in evaluating broad trends and patterns in workplace injuries; the value in assessing workplace illnesses is far more limited. It is anticipated that more advanced statistical methods will address some of the limitations confronting conventional research methods.

Reference

Schmid FA (2009), Workplace Injuries and Job Flows", ncci.com, July 31, 2009

Identifying Vulnerable Populations in Workers' Compensation Data: Limited English Proficiency Workers and Temporary Agency Workers

David Bonauto MD, SHARP Program
Washington State Department of Labor and Industries

Opportunities to use workers' compensation (WC) data for public health surveillance activities of occupational injuries and illnesses to vulnerable working populations need further exploration. Interest in using the Washington State WC data to identify possible occupational health disparities led to identification of data elements which may be used to identify vulnerable working populations. Increasing Washington State employment trends related to use of temporary agency workers and the desire for a better understanding of the occupational safety and health needs of workers with limited English proficiency led to two small descriptive epidemiologic studies focusing on these populations by the Washington State Safety and Health Assessment and Research for Prevention (SHARP) program.

Background

Limited English Proficiency workers: Limited English proficiency adults have less access to health care, low use of preventative health services, and poorer health status [DuBard, 2008]. From the relatively scant research related to occupational health disparities based on race and ethnicity, many potential factors suggest the existence of occupational health disparities based limited English proficiency. Employment distributions by industry and occupation may predispose LEP workers to more hazardous physical, chemical and biologic workplace exposures. LEP workers

Figure 1. Temporary agency and permanent status employment trends for Washington State, 1989-2005. (Source: M. Foley, Washington State Department of Employment Security, 2009).

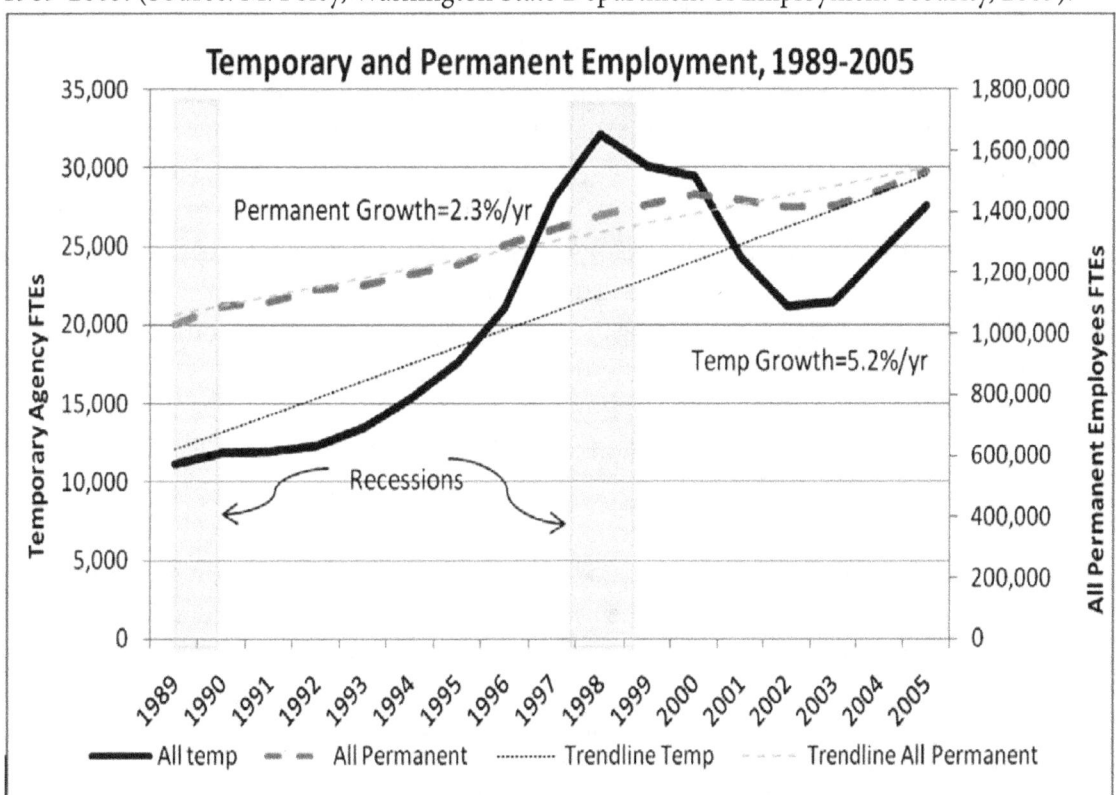

may have less occupational safety and health training, experience greater real or perceived barriers to occupational health services, have less awareness and utilization of WC insurance programs, access and use occupational health services differently, and experience worse occupational health outcomes related to longer duration of disability, less disability award amounts, and more adverse social and economic consequences of occupational injury.

WC likely can provide useful insights into the occupational injury and illness experience of LEP workers. In 2000, the United States Department of Health and Human Services Office of Civil Rights concluded that inadequate interpretation for patients with Limited English Proficiency (LEP) is a form of prohibited discrimination on the basis of national origin under Title VI of the Civil Rights Act of 1964. Compliance with this policy suggests that most WC insurers, likely collect information regarding the workers preferred language for claims communications. We conducted a descriptive epidemiology study to compare workers with LEP to those workers who were English proficient with a compensable low back disorder within Washington State WC state fund [Bonauto, 2009].

Temporary Agency Workers: Growth in temporary employment in Washington State has exceeded the growth of permanent employment during the period from 1989 to 2005. Employment pattern remain more volatile for temporary employees during periods of economic recession (Figure 1). Evidence from non-US countries suggests temporary agency workers have increased rates of occupational injury and illness when compared to employees with permanent work arrangements [Foley, 1998; Smith, 2009].

Using the Washington State unique risk classification system, we are able to compare injury rates experienced by those workers employed within temporary help agencies with those employed in permanent work arrangements. Estimates of exposure were made based on employer reported hours of work which are required as a basis for premium payment.

Methods

The Washington State Department of Labor and Industries' (L&I) State Fund (SF) is the exclusive provider of WC insurance to all Washington State employers, except those that are able to self-insured, covered by alternative WC systems (e.g., the federal government) or not mandated to have WC coverage (e.g. self-employed). Washington State statutes, regulations, and case law guide Washington WC insurance coverage requirements, claims administration procedures and insurance benefits. Both studies restrict analysis to SF data due to the absence of data elements necessary for determination of injury and illness rates (premium hours by risk classification). Data from State Fund WC claims with dates of injury from January 1, 2003 through June 30, 2006 were extracted in September 2008 from the Washington State Department of Labor & Industries (L&I) WC system. Demographic, industry and occupation of the injured claimant was derived from the WC claim record at the time of injury. Data for costs, time loss duration, benefit administration and other transactional variables were limited to a two year time period for a consistent period of claim maturity.

Limited English Proficiency workers: Limited English proficiency workers were identified by using a check box marked by the claimant that they preferred claims communications in Spanish. Cases were identified as low back disorder claims in accordance with a previously published algorithm [Silverstein, 2002]. Claim cost data reflect that which had been paid to date during the two years after the date of injury. Costs were adjusted for inflation to 2006 United States dollars according to the consumer price index. We calculated the time periods for administration of wage replacement benefits, medical benefits and authorized procedures based on date of injury or the date in which a claim was established with the SF. If benefit eligibility occurred after L&I's receipt of the claim, we calculated the number of days between the benefit eligibility date and the first provision of benefits. We used O*Net to classify occupations for the importance of trunk strength in completing the typical work performed for that occupation. The methods employed for this study are described in detail elsewhere [Bonauto, 2009].

Temporary Agency Workers: Temporary agency workers were identified from employer accounts reporting hours of employment in one of the 16 temporary help risk classifications (WIC) of the 316 WIC classifications used by Washington State to assess WC premiums. Individual WC claims are assigned a WIC from the employer's assigned WICs and the type of work the claimant was performing at the time of injury. We developed a set of comparison WIC between the temporary WIC to those with standard employment WIC to assess injury rates between the two groups by type of work. The methods employed for both studies are described in detail elsewhere [Foley, 1998; Smith, 2009].

Figure 2. Median medical and total claim costs for compensable low back disorder claims by English proficiency, Washington State state fund, January 2003 – June 30, 2006. [Source: Bonauto, 2009]

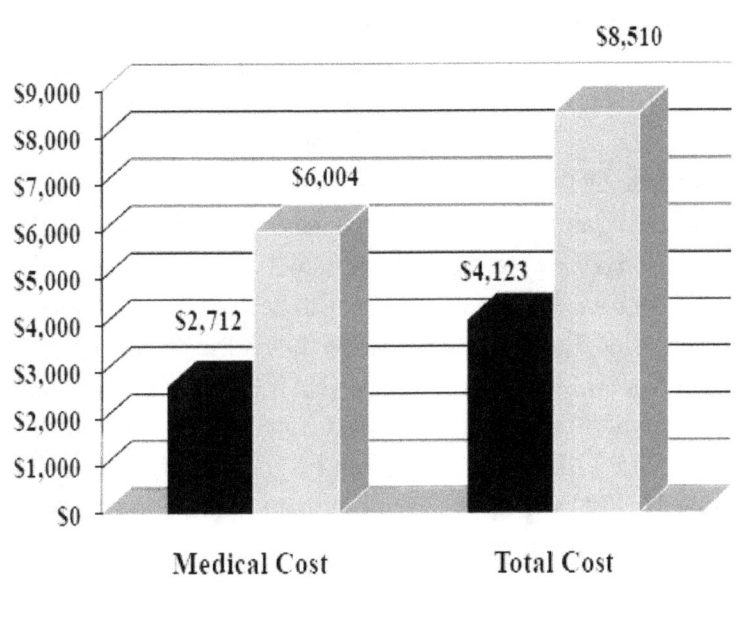

* Costs indexed 2005 CPI at 2 years common claim maturity from date of claim receipt; actuary incurred estimates; Differences are statistically significant – (p<0.0001)

Figure 3. Claims rate comparisons for matched risk classifications between temporary agency employment and standard employment arrangements; Washington State state fund, 2001 – 2005.

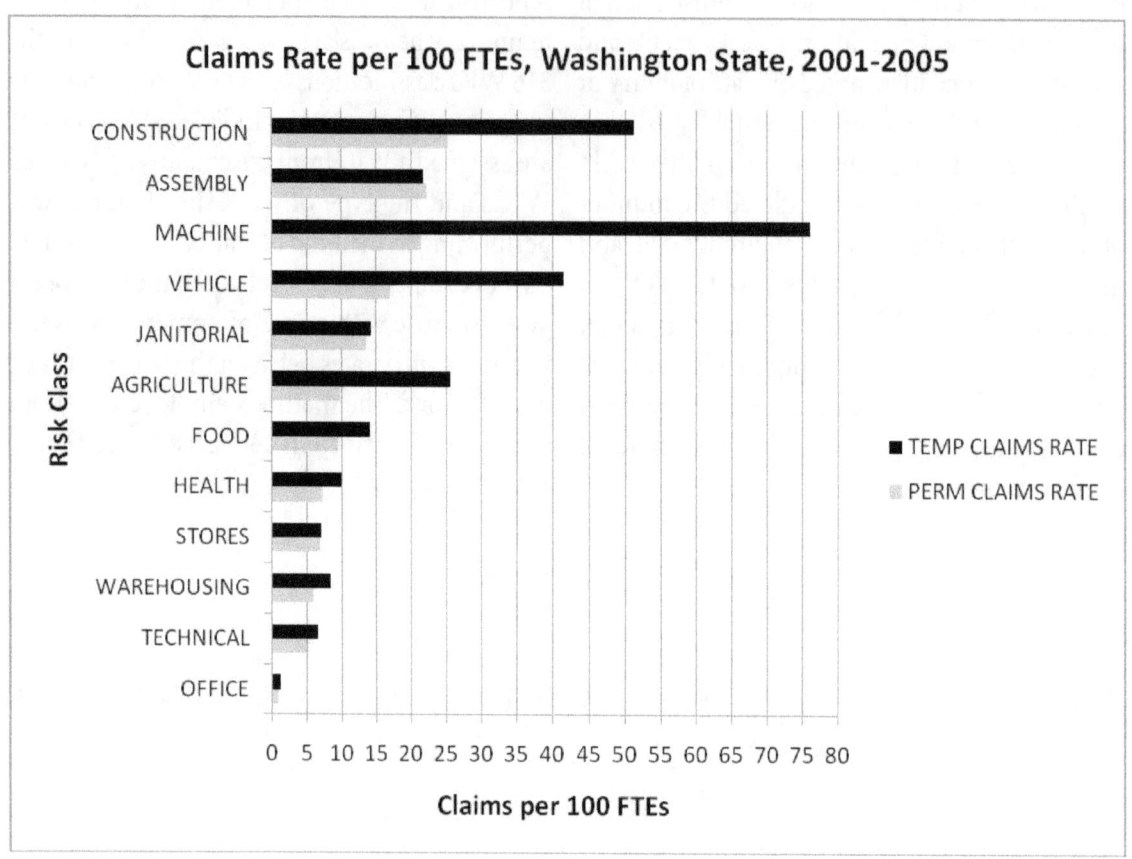

Results

Limited English Proficiency Workers: Of the 20,805 Spanish language preferring (SLP) claimants for the time period under study, 2,266 (10.9%) were claims filed for non-traumatic low back disorders (LBD). During the same time period, there were 449,734 English language preference (ELP) claims of which 57,069 (12.7%) were for non-traumatic LBD. A greater proportion of SLP LBD claims filed were accepted and resulted in lost work time than ELP LBD claims. There were significant differences in the demographic, employment, and occupational characteristics between the SLP and ELP compensable claimant populations. SLP claimants were more likely to be male, less than 35 years old, married, have at least one child, live in an economically distressed county, be overweight but less likely to be obese, employed less than one year, employed in company with 11-50 FTE and have an income less than $25,000. The employment distribution of all LBD compensable claimants differed with the SLP claimants appearing to be more employed in the following NAICS sectors: Agriculture, forestry, fishing, and hunting (NAICS Sector 11); Construction (23); Manufacturing (31-33); Administrative, Waste Management and Employment Services (56); and Accommodation and Food Services (72) than ELP LBD claimants. The SLP LBD compensable claimants had greater time loss duration, greater medical and total claim costs, more use of physical therapy and vocational services than the ELP LBD compensable claimants. With the exception of the timeliness for providing the first time loss payment, the time periods for provision of insurance benefits did not differ between the SLP and

ELP populations. SLP compensable claimants received less back surgery and had comparable permanent partial disability payments to the ELP population. Employers were more likely to protest the acceptance of a SLP compensable than one in an ELP LBD compensable claim.

Temporary Agency Workers: When comparing injury rates between comparable WIC classifications between temporary agency workers and workers in standard employment arrangements, there are elevated rates in the temporary risk classifications associated with agricultural operations, commercial vehicle operations, machine operations and construction laborers (Figure 3).

Discussion

Exploring data elements that identify vulnerable workers within WC data may be useful for public health surveillance. Our use of Washington State's exclusive state WC system and data elements identifying limited English proficiency workers and temporary agency workers provide information to characterize possibly increased occupational injury rates and poor occupational health outcomes. While the results are subject to possible limitations related to differential reporting of occupational injuries and illnesses by vulnerable groups, variation in the knowledge, attitudes and beliefs associated with workers compensation, and other limitations; the results provide many hypotheses for additional research.

Acknowledgements: Washington State Department of Labor and Industries SHARP staff, including Darrin Adams, David Bonauto, Joyce Fan, Michael Foley, Barbara Silverstein, and Caroline Smith conducted these studies. The information presented above resulted from descriptive epidemiology studies funded by Washington State Department of Labor and Industries. The content and opinions are solely the responsibility of the authors and may not represent the opinions and policies of the Washington State Department of Labor and Industries.

References

Bonauto DK, Smith CK, Adams DA, Fan ZJ, Silverstein BA, Foley MP. Language preference and non-traumatic low back disorders in Washington state workers' compensation. Am J Ind Med. 2009 Aug 31. [Epub ahead of print]

Dubard CA, Gizlice Z. Language spoken and differences in health status, access to care, and receipt of preventive services among US Hispanics. Am J Public Health. 2008. 98:2021-2028.

Foley MP. Flexible work, hazardous work: the impact of temporary work arrangements on occupational safety and health in Washington State, 1991-1996. Research in Human Capital and Development. 1998. 12: 123-147.

Silverstein B, Viikari-Juntura E, Kalat J. Use of a prevention index to identify industries at high risk for work-related musculoskeletal disorders of the neck, back, and upper extremity in Washington state, 1990-1998. Am J Ind Med. 2002 41: 149-69.

Smith CK, Silverstein BA, Bonauto DK, Adams D, Fan ZJ. Temporary workers in Washington State. Am J Ind Med. 2009 Jul 17. [Epub ahead of print]

How to Make Interventions Work: An Insurance Perspective

Héctor Upegui[1], Victor Schultheiss[2] Centre of Competence for Workers' Compensation Insurance, Munich Re

Prevention does not have the same meaning for all who make decisions on investing in it. For example, decision makers who focus on occupational safety and health (OSH), understand and apply prevention in different ways than decision makers, who focus on insurance. In this paper we will show some of our findings on the differences in the perspectives of decision makers in OSH and insurance, in order to demonstrate how the underlying differences in perspective lead to different actions, given the same data. Understanding these differences, might help to guide decision makers from both OSH and insurance to better align their efforts in order to benefit workers, employers, and insurers.

In this article prevention should be mainly understood as primary prevention in occupational safety and health, and compensation as all those benefits in kind and in cash provided by a workers' compensation system in case there is an occupational accident (OA) or an occupational disease (OD). Terminology should also be applied to any system without any distinction whether it is purely public or with private participation.

Compensation and primary prevention work together in what we consider as "The basic unit of the traditional working environment" (Figure 1), but with emphasis on different aspects. Although everything is about occupational hazards and its impact in the whole, as well as beyond the

Figure 1. Relationships among Workers' Compensation Insurance (WCI), Occupational Safety and Health (OSH), and Occupational Hazards in the Traditional Work Environment

WCI: Workers' Compensation Insurance
OSH: Occupational Safety and Health

Figure 2. Each system collects and manages information in a completely different way

The basic unit and collecting of data

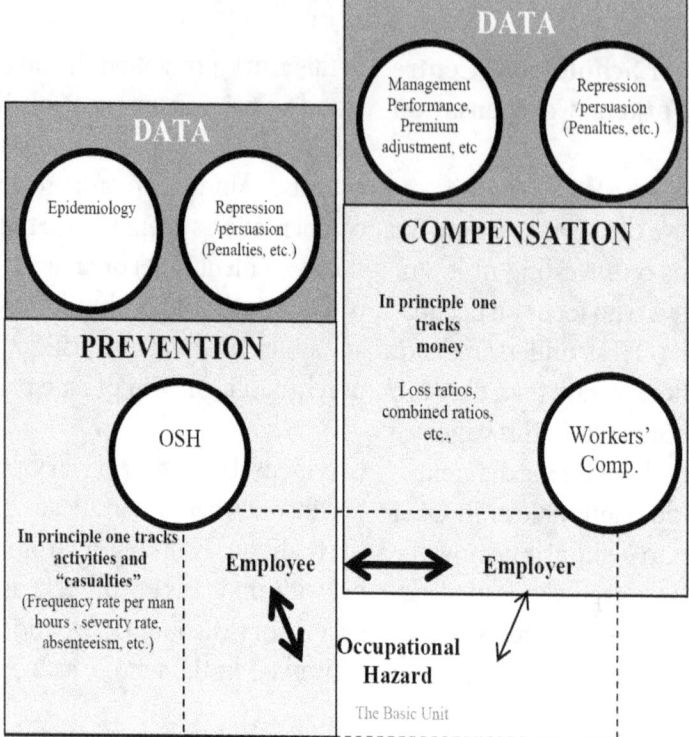

Figure 3. Theoretical correlation of prevention and premiums in workers' compensation insurance. Increasing prevention ought to be accompanied by lower premiums.

WCI vs. OSH

The employer's dilemma

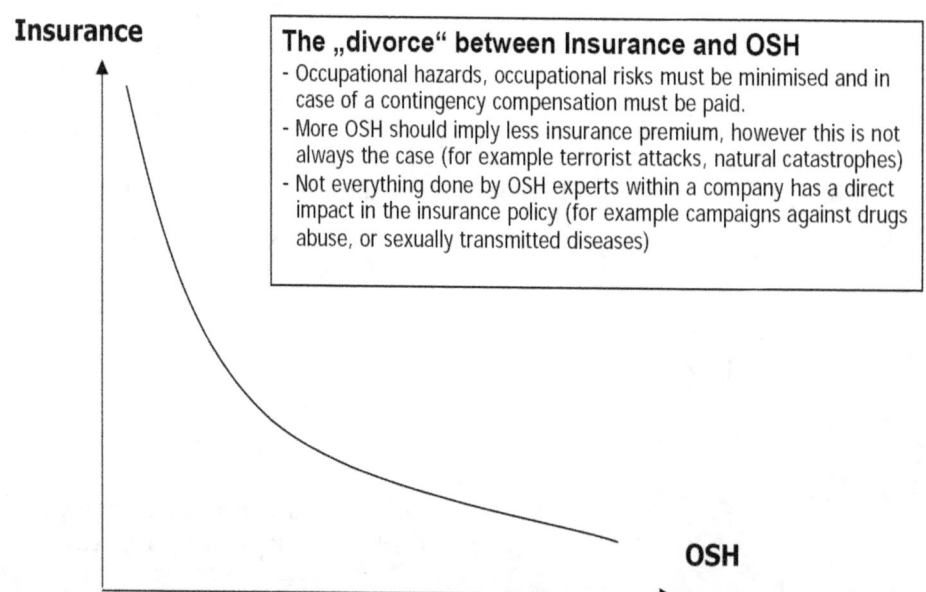

"Basic unit" (families, society, productivity, competitiveness, etc.), revising the origins of primary prevention and compensation they both differ from each other. Primary prevention grew out of the observation that there was a need to protect the person from undesirable effects of occupational hazards, while compensation answered to the interest of governments/societies to guarantee a fair compensation for the victim in case of an OA or an OD. In other words, while the first one focused on the interactions of human beings and hazards, the second one developed mechanisms to allow employers to transfer the responsibility of payments to an institution (Workers' Compensation Insurance).

There is no doubt about the need for cooperation between compensation and prevention to better achieve respective and common objectives. However a mismatch can occur when trying to put the two systems together without understanding the different bias of each system. For instance, each accident or disease is a failure for primary prevention, but is an input for compensation. While the main framework for prevention is that accidents/diseases do not happen, the conceptual approach for compensation is that accidents/disease will happen. Compensation is about building up capitals to pay out future expected losses, while primary prevention is about using all possible resources to avoid future losses. Primary prevention believes in zero accidents, while compensation always includes in its calculations a deviation from expected losses.

Prevention counts with best scenarios, compensation conversely works with worse scenarios. In primary prevention the "clients" are employees/workers, while in compensation all of them are insureds or victims. For the employee/worker primary prevention is about activities, while compensation is about money. Compensation has restrictions in order to guarantee that real victims receive what they are entitled to; on the contrary, primary prevention is in favor of widening its scope of coverage. In primary prevention the system looks for evidence in order to fulfill its objectives, however the user is in principle passive regarding its interaction with the system, but in compensation, as the insured has to claim, another dynamic exists. In primary prevention it is quite often to find that the system provides/offers more that what the employee/worker takes, while in compensation is normally the other way round.

Another important characteristic that heavily influences information systems under "The Basic Unit" is the relationship between the participants in each system. In primary prevention, the relationship between the employer and the employee is person to person (P2P), whereas for compensation the relation between the employer and the workers' compensation institution is Business to Business (B2B).

For the above reasons, each system collects and manages information in a completely different way (Figure 2). Trying to make both systems work similarly as to how they handle information is counterproductive, expensive and unnecessary. Information under primary prevention is mainly epidemiologically driven, whereas in compensation is financially driven. For instance, prevention registers incidents and accidents when they happened. Compensation registers these accidents whenever they are reported or claimed to the insurer, or even more when they are already paid.

Should primary prevention be part of compensation?

There has been always a discussion as to how much primary prevention should be included in compensation (for instance that institutions dealing with compensation should be responsible also for pre-employment examinations, or for periodical screening of ODs, etc.). This discussion also brings in additional elements to prove differences when discussing information systems.

Figure 3 shows how premiums and prevention should ideally correlate. One would expect that increasing prevention should lead to decreasing premiums and diminishing prevention to increasing premiums.

This diagram does not reflect reality, however, because compensation insures what is defined in the law (workers' compensation act, employment accident act, etc,). In some countries/jurisdictions for instance, terrorism and/or natural catastrophes are covered under workers' compensation insurance. In these cases claims' frequencies do not change in proportion to the preventive measures taken, because the room employers have for maneuvering is too limited. In contrast to other occupational hazards such as noise or the effects of chemical substances, these risks are not subject to an employers' influence. On the other hand, not everything that a company does on primary prevention within the scope of occupational safety and health has direct effects on workers' compensation insurance, even though it does indeed affect the health of the employees (e.g. campaigns to combat sexually transmitted diseases).

Furthermore, primary prevention deals with frequency and severity. Compensation is heavily driven by severity, and therefore its repeated actions to curtail costs through claims handling (secondary and tertiary prevention) strategies.

These are just some reasons to propose that compensation is better for secondary and tertiary prevention than for primary prevention.

Conclusion

To prevent is better than to heal, but what to do if prevention fails? On the other hand, compensation needs primary prevention: What measures can be taken in order to maintain a healthy compensation system? There should be, of course, cooperation for the profit of all parties involved. In designing this mutual cooperation, understanding the respective restrictions might be useful. One possible way to save time and resources might be that each party take as input what the other one already has as output, while considering that compensation is better for secondary and tertiary prevention than for primary prevention.

Meanings are not always the same because the way in which each actor understand and assumes its role is completely different. It is not about defining who does it better, is about bringing together strengths in order to reach common objectives.

[1] Physician, Master in Social Security, specialized in OSH Management, Senior Consultant, Centre of Competence for Workers' Compensation Insurance, Munich Re
[2] Managerial Economist, Head of Unit, Centre of Competence for Workers' Compensation Insurance, Munich Re

Narrative to Accompany "Barriers to Reporting"

Lenore S. Azaroff ScD, University of Massachusetts at Lowell

A growing body of literature continues to confirm workers' and advocates' experiences that some medical treatment and lost work time required by work-related injuries and illnesses are not covered by workers' compensation insurance (WC) (1-7). Recent state supplements to the Behavioral Risk Factor Surveillance System national telephone survey (BRFSS) asked sub-sets of respondents about their self-reported work-related injuries (avoiding the question of potentially more ambiguous work-related illnesses) and health care for those injuries. Results from Washington State in 2002 showed that 52% of 321 respondents who reported work-related injuries said that the care for those injuries was covered by WC (8), consistent with less than 60% of 110 such people in the 2007 Massachusetts BRFSS (9). A 2008 survey of 4,387 workers in low-wage industries in Chicago, Los Angeles, and New York City found that 12% of respondents reported having a workplace injury requiring medical attention during the previous three years, but that care for just six percent of these injuries was covered by WC (10).

In considering barriers to reporting work-related injuries and illnesses to workers' compensation (WC), it is useful first to consider the conditions that would have to be in place for workers to report. Logically, there are two sets of conditions that would facilitate workers reporting their injuries:

1. a worksite system in place to support reporting, and no strong disincentives to using the system, or

2. workers taking initiative by themselves to initiate reporting, and presence of strong incentives to take this initiative

Worksite System in Place to Support Reporting, and No Strong Disincentives to Using the System

What types of employment arrangements lack this first set of conditions? An immediately obvious example is found in the informal sector: day labor, under-the-table, or informal employment, in which basic systems for wages, taxes, and insurance are lacking or incomplete. Employers may fail to recognize employees as working for them at all, or might misclassify them as independent contractors. This type of work is unlikely to include systems to inform workers about procedures for reporting as well as WC coverage for all employees, administrative staff and procedures to process claims, and other necessary components. Inadequacy of such systems is also probable in legal but non-traditional systems of indirect employment such as contracting and employment through temporary agencies.

However, systems for reporting injuries and illnesses to WC can break down in employment arrangements that are legal, even traditional, but simply decentralized. In the author's personal experience, a major janitorial services company with many thousands of employees maintained scrupulous, detailed policies related to health and safety and workers' compensation at their central offices, but this information did not typically reach this company's employees, either supervisors or cleaning workers, at worksites literally blocks away. At that time, workers relied on coverage by the Massachusetts Uncompensated Care Pool ("Free Care") or, rarely, other forms of insurance, to treat injuries and illnesses caused by work. They took for granted that missing too many days of work due to such health problems would result in losing their jobs.

This type of decentralized work is common in industries such as construction, landscaping, cleaning, and home health and personal services, where workers have little direct contact with high-level managers or company policies. A recent in-depth study of personal assistance services workers in California compared WC experience of those working for an agency versus independent providers. Both groups are entitled to WC by law. However, the independent providers' access to WC was hampered by lack of a central employer system to inform them about WC and to help them file claims smoothly. This group of workers reported "the runaround," i.e., months of delays, lack of correct contacts, and incorrect information, which in turn sometimes resulted in delayed medical treatment, deteriorating health, and loss of income (11).

Failure to systematically transmit information about WC in general and an employer's WC policies and procedures in particular is a common barrier to reporting at both informal and decentralized worksites. Focus groups and interviews with low-wage workers, organizations, and employers in California have revealed a widespread lack of understanding that chronic pain, non-acute injuries, are work-related and serious enough to report (12). Of 168 hotel room cleaners with work-related injury not reported to WC, 18% explained "I did not know how" (13). The most common reason given for not filing for WC in the Washington 2002 BRFSS was that respondents "did not know that they could file" (8).

The lack of reporting systems for many workers in low-wage industries is revealed by findings in the three-city survey of 4,387 workers:
> Fully 43 percent of seriously injured respondents reported that they were required to work despite their injury; an additional 30 percent said their employer refused to help them with the injury; 13 percent were fired shortly after the injury; 10 percent said their employer made them come into work and sit around all day; 4 percent were threatened with deportation or notification of immigration authorities; and 3 percent were told by their employers not to file a workers' compensation claim. Only 8 percent of employers instructed injured workers to file a workers' compensation claim (10).

These and other illustrations of lack of adequate systems were provided in the focus groups and interviews in California low-wage industries. They found that sub-acute injury and pain are so common that they are considered a normal part of the job, to the point that people continue to work despite nearly unendurable pain because they believe they have no alternatives; some employers attempt to make workers pay for their own medical care; some employers deny that injured workers are their employees; and some claim to lose injury reports, withhold WC documentation, or even mislead workers about WC until the statute of limitation for filing expires (12).

This brings us to the second half of this first set of conditions: *and no strong disincentives to using the system*. In some cases worksite systems are in fact in place, but barriers exist in the form of powerful disincentives to use the system. Fear of job loss and retaliation was the most commonly reported concern about reporting to WC in Lashuay and Harrison (2006), and the most common reasons for not filing for WC in the Washington (8) and Massachusetts (9) BRFSS surveys. Studies of immigrant workers, in particular, suggest that expectation of retaliation and/or involuntary job loss is an assumed aspect of the filing experience (14).

Far from reflecting a lack of a system, Lashuay and Harrison (2006) found that for many, retaliation is the system: "Several supervisors reported that firing employees who complained or filed workers' compensation claims was company policy."

Adding a dimension to fear of retaliation against oneself by one employer, some employees, particularly those working through employment agencies or hiring halls, report concern about being blacklisted by future potential employers and about being ostracized by fellow workers. Main reasons cited by personal assistance workers for unwillingness to report to WC included fear of job loss, fear of damaging their reputation for future jobs, and their commitment to the consumers of their services (11).

Fear of job loss may be complemented by fear of loss of rewards, not just for the worker, but for an entire peer group, due to the particular safety incentive programs that reward workdays without reported injuries. One commercial supplier of this type of incentive program describes it this way:

> The game begins after all employees have been assigned to a working team. Teams are normally 50 to 75 employees who see and talk to one another on a regular basis. Each employee is then given a B-Safe Safety Bingo Card, and the game is played like regular Bingo. To win, 5 numbers in a row must be blacked out. Each working day a B-Safe Safety Bingo number is drawn. Daily numbers are posted in a visible place like the lunch room or on employee bulletin boards. The game ends for all employees on a specific "working team" when one of the employees sustains an occupational injury. All other employees continue to play until a pre-determined number of winners come forward. (www.blogcatalog.com/topic/safety+incentives/ accessed September 14, 2009).

Little research is available on the effect of such programs on injury reporting behaviors. Anecdotally, and in the author's experience, they can provide a powerful disincentive to using WC. They are also becoming common. The 2008 Massachusetts Worksite Health Improvement Survey included the question "Does your company reward employees (e.g. paid time off, bonuses, prizes, etc.) for periods of time when there are no injuries reported?" Of 890 employer respondents, eight percent responded "yes." These included 18% of respondents in manufacturing, 16% of respondents in construction, and 10% in transportation and warehousing (15), all sectors with overall high rates of occupational injuries.

Workers Taking Initiative by Themselves to Initiate Reporting, and Presence of Strong Incentives to Take this Initiative

Workers can report their injuries and illnesses to WC directly without participating in a worksite system that supports WC use. For example, explaining the work-relatedness of a condition to a health care provider can begin the process of WC coverage of health care, and reporting to an insurer or state agency can result in replacement of lost wages. This, however, presupposes knowledge and motivation on the part of the injured worker.

Many people are simply not familiar with WC. A Massachusetts Department of Public Health Survey of 1,428 working patients at five community health centers found that 39% of patients had never heard of WC. These included 50% of service workers and 47% of operators, fabricators, and laborers surveyed (16). A household survey of 160 Southeast Asian immigrants and refugees in Lowell, Massachusetts found that 69% responded "No" or "Don't know" when asked whether they had heard of WC after hearing a brief description of the system (17).

If people are familiar with the WC system, what would motivate them to take the initiative to file claims regardless of their worksite practices? An obvious potential incentive would be access to health care. Some researchers in certain low-wage settings, however, have found that reporting injuries leads to trivialization of injuries by company doctors and nurses, access to only token treatment, and still being forced to work injured (12, 18).

Sometimes pursuing a claim through WC rather than other insurers can result in delays in treatment. A study of California WC cases for non-specific low back pain examined administrative delay for 35,304 cases from 1993 through 2000. This found delays of more than two weeks for 30% of cases, and delays of greater than 90 days for 2,066 (five percent) of cases. For those with the least severe injuries, just two weeks of administrative and treatment delays were associated with a 77% greater chance of chronic disability (19).

Given these realities, it seems at times to be a rational choice to use other forms of coverage, even for people who are familiar with WC. Of 941 hotel room workers surveyed, 35% reported at least one work-related injury to WC, and 168 (18%) had a work-related injury that they did not report to WC. Of those who reported to WC, 54% said their claim was denied. Of those who did not report to WC, 43% said "It would be too much trouble." Some did not report because they thought that the injury would get better, believed the injury was not covered by any insurance, or did not want to "lose work time" (13).

These findings about hotel workers are similar to 2000 findings by Rosenman among, not day laborers, but mostly unionized autoworkers employed at large central plants. This illustrates the need for a strong incentive for people with other options. If fact, the Massachusetts BRFSS found that of people with work-related injuries earning less than $50,000, 71% reported that their care was covered by WC, while of people earning over $50,000, just 47% did. The most frequent other source of payment reported (24%) was private insurance. A common reason given for not reporting to WC was that it was "easier to use other insurance" (9).

References

1. Biddle J, Roberts K, Rosenman KD, Welch EM. What percentage of workers with work-related illnesses receive workers' compensation benefits? J Occup Environ Med. 1998 Apr;40(4):325-31.

2. Boden LI, Ozonoff A. Capture-recapture estimates of nonfatal workplace injuries and illnesses. Ann Epidemiol. 2008 Jun;18(6):500-6. Epub 2008 Feb 20

3. Leigh JP, Robbins JA. Occupational disease and workers' compensation: coverage, costs, and consequences. Milbank Q. 2004;82(4):689-721.

4. Morse T, Dillon C, Kenta-Bibi E, Weber J, Diva U, Warren N, Grey M. Trends in work-related musculoskeletal disorder reports by year, type, and industrial sector: a capture-recapture analysis. Am J Ind Med. 2005 Jul;48(1):40-9.

5. Morse T, Dillon C, Warren N. Reporting of work-related musculoskeletal disorder (MSD) to workers compensation. New Solut. 2000;10(3):281-92.

6. Rosenman KD, Gardiner JC, Wang J, Biddle J, Hogan A, Reilly MJ, Roberts K, Welch E. Why most workers with occupational repetitive trauma do not file for workers' compensation. J Occup Environ Med. 2000 Jan;42(1):25-34.

7. Lipscomb HJ, Dement JM, Silverstein B, Cameron W, Glazner JE. Who is paying the bills? Health care costs for musculoskeletal back disorders, Washington State Union Carpenters, 1989-2003. J Occup Environ Med. 2009 Oct;51(10):1185-92.

8. Fan ZJ, Bonauto DK, Foley MP, Silverstein BA. Underreporting of work-related injury or illness to workers' compensation: individual and industry factors. J Occup Environ Med. 2006;48:914-22.

9. Massachusetts Department of Public Health Unpublished Data, 2009.

10. Bernhardt A, Milkman R, Theodore N, et al. Broken laws, unprotected workers: violations of employment and labor law in America's cities. Center for Urban Economic Development, National Employment Law Project, UCLA Institute for Research on Labor and Employment, 2009.

11. Scherzer T, Wolfe N. Barriers to workers' compensation and medical care for injured personal assistance services workers. Home Health Care Services Quarterly. 2008;27(1):37-58.

12. Lashuay N, Harrison R. Barriers to Occupational Health Services for Low-Wage Workers in California. San Francisco: University of California, April 2006.

13. Scherzer T, Rugulies R, Krause N. Work-related pain and injury and barriers to workers' compensation among Las Vegas hotel room cleaners. Am J Public Health. 2005;95:483–488.

14. deCastro AB, Fujishiro K, Sweitzer E, Oliva J. How immigrant workers experience workplace problems: a qualitative study. Arch Environ Occup Health 2006;61(6):249-258.

15. Massachusetts Department of Public Health Unpublished Data, 2009.

16. Massachusetts Occupational Health Surveillance Program. Occupational health and community health center (CHC) patients: a report on a survey conducted at five Massachusetts CHCs. Boston: Massachusetts Department of Public Health, April 2007.

17. Azaroff LS, Levenstein C, Wegman DH. The occupational health of Southeast Asians in Lowell: a descriptive study. Int J Occup Environ Health 2004;10:47–54.

18. Marin AJ, Grzywacz JG, Arcury TA, Carrillo L, Coates ML, Quandt SA. Evidence of organizational injustice in poultry processing plants: possible effects on occupational health and safety among Latino workers in North Carolina. Am J Ind Med. 2009; 52:37–48.

19. Sinnott P. Administrative delays and chronic disability in patients with acute occupational low back injury. J Occup Environ Med 2009;51:690-699.

Comparing Lost Work Days under Workers' Compensation and Short-term Disability, Evidence from IBI's Disability Benchmarking Data

Brian Gifford, Thomas Parry, William Molmen and Kim Jinnett, Integrated Benefits Institute

Background

Employers are looking for every opportunity to improve business results during these difficult economic times. One important strategy is to ensure that the workforce is healthy, at work and engaged for optimal productivity. This means addressing more than just incidental sick days and presenteeism: disability lost-time is also a potential area for productivity improvements.

However, the programs for managing disability absences – workers' compensation (WC) for occupational injuries/illnesses, and short-term disability (STD) for non-occupational conditions – have vastly different design features, and are often managed in separate administrative silos. WC is a statutory program that integrates medical care and wage replacement payments; elimination periods are standardized according to state law; medical care is extended until a point of maximum medical improvement and temporary total disability benefits seldom are capped; it has a permanent disability component to compensation that is often not based on actual loss of earnings (for this reason, litigation is not infrequent).

By contrast, STD is a contractual system between the employer and employee with a maximum period of benefits, has an elimination period to qualify for benefits set by the contract, typically does not integrate medical care as part of the program and is litigated relatively infrequently. Thus, because of these design differences, workers with similar conditions can have very different durations of absence depending on whether or not their injury is work-related.

The implication for employers is clear – identifying the disability management strategy that most effectively returns employees to work, and adopting that strategy whether or not an employee's condition is work-related, could result in lost-productivity savings. What is not clear is whether WC programs return employees to work more quickly than STD programs, or vice versa. Employers typically lack even the most basic information about how lost time differs across the two programs.

Data and Analysis

To better understand differences in disability durations for incidents that are nominally similar, we compared calendar year 2007 lost work days for WC and STD claims contained in the Integrated Benefits Institute's (IBI's) benchmarking database. Comparing WC and STD claims is challenging because the program characteristics are so different, but also because STD programs address a wider diversity of disabling health conditions than WC programs.

For these reasons, we attempt to take out the effect of program differences by limiting our analysis to a subset of claims with the following characteristics: (1) the disability incident occurred in January 2007, (2) benefits commenced within one week of the incident, (3) STD claims had a maximum benefit duration of at least one year, and (4) the claim diagnoses were for conditions that are typical of WC incidents. (predominately injuries and musculoskeletal conditions such as back disorders, joint derangement, muscle and tendon disorders, back sprains and strains, contusions, fractures, joint sprains and strains, and open wounds.)

While claims with these characteristics are not necessarily representative of the universe of STD or WC claims, they nonetheless provide a basis on which lost work days can be reasonably compared. Our final benchmarking sample consists of 1,690 STD claims, and 871 WC claims.[1]

In our sample, at the 25th, 50th (median), and 75 percentiles, the distribution of lost work days for WC and STD claims are statistically identical. Overall, however, WC claims have a mean of 57 lost work days, compared to 49 days for STD claims. While this difference is statistically significant, as the figure below shows, it largely reflects the influence of WC claims with the longest durations, as the figure below shows. At the 90th percentile the duration of WC claims is 45% longer than the duration of STD claims.

Commentary

The evidence suggests that WC and STD claims with roughly similar claim characteristics also have similar lost-time outcomes – but only up to a point. Understanding why the longest WC claims have more lost work days than the longest STD claims will likely require more analysis. While severity almost certainly plays a role in the duration of absence within either program, the data provide little information to explain why the longest WC conditions would be systematically more severe than the longest STD conditions.

What's more, within the WC group of claims, the role of litigation – as both a cause and effect of lost work time – deserves a closer look, with particular focus on how it overlaps with case management factors that are under employers' control. In our sample, WC claims in the top 10% of lost work days were almost 1-1/2 times more likely to report attorney involvement than WC claims in the bottom 90% (28%, compared to 11%). Interestingly, there were no significant differences in the percentage of claims that resulted in permanent disability awards (although some of these claims may result in permanent disability awards at a later date).

Further analysis may find that there is no difference in medical severity across WC and STD, despite the longer durations for WC at the upper end. This may tell us that the unique characteristics of the WC system are extending disability for some complex claims. Early IBI research (1) demonstrated that a "sports medicine" approach to medical treatment of injury – both in the amounts and timing of medical treatment – affects time away from work. An employer focus on return to work (RTW) also can produce savings and superior RTW results apart from the effects of medical treatment. To the extent that WC cases at and above the 90th percentile of lost work days could benefit from an intensive approach to medical treatment, the early reporting and acceptance of a claim by an employer could enhance the ability to provide treatment and management appropriate to an early return to work.

In addition, IBI case studies (2) and research (3) demonstrate the value to employers and employees (in enhanced satisfaction and reduced litigation)

---

(1) Limiting claims to these criteria not only provide the most comparable WC and STD cases that the data allow, they also provide the best opportunity to observe the full extent of lost work days. The January criterion provides the maximum opportunity for claims to resolve within the one-year time frame. Although 61% of WC claims were classified as "open" by data providers, 95% had fewer lost work days than would be required for the claim to roll-over into the next calendar year. By comparison, 93% of STD claims were classified as closed, and 97% had fewer lost work days than would be required to roll-over into the next calendar year. Because we are looking only at calendar year 2007 lost work days, the one-year benefit duration criterion effectively excludes STD cases that were arbitrarily closed because the maximum duration had been reached (which can result in a transition to LTD). For these reasons, claims that meet all requirements represent only a subset of the January claims in the Benchmarking data. Almost 75% of January WC TTD claims were excluded because they did not have an ICD-9 code by which they could be compared to STD claims. Applying all criteria for inclusion, analyzed WC claims represent 18% of the January WC Benchmarking claims, while analyzed STD claims represent only three percent of the January STD Benchmarking claims.

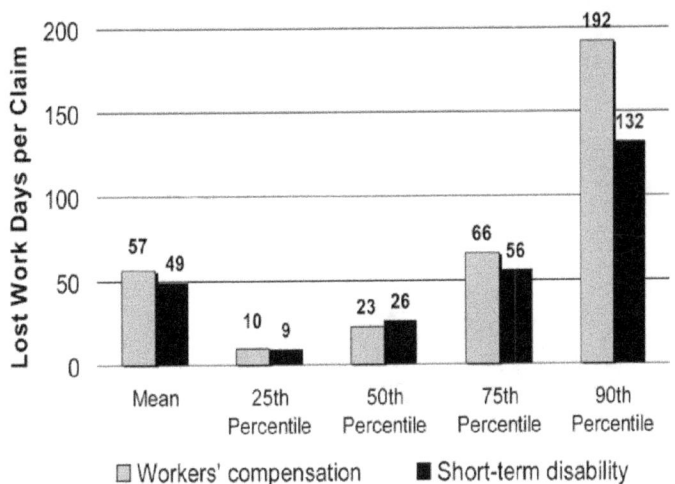

in having a single contact for early reporting by employees of all disability conditions. There is demonstrated value in commencing medical treatment and wage replacement benefits under such an approach without first determining which program is responsible, thus delaying case management, benefits and, sometimes, medical treatment.

Unfortunately, IBI Benchmarking data do not contain the type of information required for such analysis. Nor is it clear how many employers currently link case management and treatment information back to claims information, then subject these data to rigorous analysis. Nonetheless, the comparative findings of this Quick Study represent a first step towards framing the kinds of questions that can fruitfully guide future analyses and employer interventions.

References

1. "Return To Productivity," Integrated Benefits Institute, September 1996. http://ibiweb.org/do/PublicAccess?documentId=507

2. "Getting to Benefits Integration: Lessons from Employers that Implemented Health & Productivity Management," Integrated Benefits Institute, July 2001. http://ibiweb.org/do/PublicAccess?documentId=479

3. "How Employers Look at Integrating Health & Productivity Management: A Survey of Integrated Benefits Best Practices," Integrated Benefits Institute, July 2002. http://ibiweb.org/do/PublicAccess?documentId=493

Linking Workers' Compensation and Employment Security Data for Occupational Health and Safety Surveillance

Michael Foley, Washington State Department of Labor and Industries

Opportunities to extend our understanding of predictors of occupational injury and illness using both national surveys such as the Health and Retirement Survey, the Behavioral Risk Factors Surveillance System, employer-level records of the Survey of Occupational Injury and Illness and administrative databases such as workers' compensation claims records and hospital discharge databases have been suggested and implemented by researchers (Reville, 2001; Boden, 2008). One of the largest and most potentially useful of these administrative databases is the Quarterly Covered Employment and Wages (QCEW) program, formerly referred to as the ES-202 program (BLS, 2009).

Background

This cooperative program involving the Bureau of Labor Statistics (BLS) of the U.S. Department of Labor and the State Employment Security Agencies (SESAs) produces a tabulation of employment and wage records for workers covered by state unemployment insurance (UI) laws and federal workers covered by the Unemployment Compensation for Federal Employees (UCFE) program.

The QCEW program derives its data from quarterly tax reports submitted to State Employment Security Agencies. These reports provide information on the number of people employed and the wages paid to the employees each quarter. The program also obtains information on the physical location and industrial activity of each reported establishment, and assigns location and industrial classification codes accordingly. Unemployment insurance coverage is broad and coverage is basically similar from state to state, making this database especially useful for making state comparisons.

Certain industry exclusions should be taken into account when using this database. Wage and salary agricultural employees are not fully covered, nor are self-employed farmers and self-employed nonagricultural workers, or certain domestic workers. Other exclusions include members of the military, workers covered by the railroad unemployment insurance system, and certain state and local government workers.

Under the federal requirements of the SESA unemployment insurance program, employers with covered employees must register with their state security agency. Information collected from the business includes legal name, physical address and phone number, ownership type (private, non-profit, public), industry type, multiple location status and FEIN. Each month employers record the total number of employees working in the pay period containing the 12th of each month and the total compensation paid during each calendar quarter. For multiple location businesses a separate record is maintained for each establishment. These data are maintained in the Employer database. The NAICS industry (previously SIC) assignment of each employer is audited and updated once every three years. In addition to the employer-level data, the SESA program receives the individual wage and hours worked records for each covered employee, along with their full name and SSN. These are reported on a quarterly basis and maintained in the Wage database. Each individual record also includes the employer UI account number, which provides a linkage to the Employer database.

Researchers may be granted access to the individual employer records of the QCEW program by submitting a research proposal to the Bureau of Labor Statistics (see http://www.bls.gov/bls/blsresda.htm). Access may also be obtained through data-sharing agreements between state agencies.

The properties of the databases within the QCEW program make it particularly useful for researchers interested in gaining a fuller understanding of the work setting in which injuries and illnesses occur and to investigate the long-term consequences of injuries. There are several ways in which linkages between administrative databases in workers' compensation systems and the databases maintained by state unemployment insurance systems have been used in recent and on-going research projects. These are reviewed below and some future directions for this research are proposed.

Using SESA Data for Industry-Based Surveillance

The SHARP Program at the Washington State Department of Labor and Industries is conducting an industry-based surveillance project to reduce injuries in the trucking industry. The first phase of this project was to conduct surveys with both long-haul and less-than-load short haul drivers and their employers to gather information on exposures, potential solutions and barriers faced. To better understand the distribution of responses across industry subsectors and company sizes, we wanted also to obtain information about the businesses for which each driver worked and then link that administrative data to the survey response record. The scope of the study included all active businesses with at least five employees in the set of 4-digit NAICS industries covering the for-hire trucking industry. To select eligible employers we extracted the following data: UI account number, Universal Business Identifier (UBI), name of business, address and phone, six-digit NAICS code, county code, count of employees and total paid compensation. Using the UBI number as a link, we merged this file with workers' compensation claims data for the most recent five-year period. We constructed a claims rate history for each employer, including type of claim and total lost work days, along with the SESA-derived data. From the Wage file within the Employment Security database we used the UI account number to extract individual employee hours and wages records for each employer for the previous five year periods. These were used to establish eligibility for inclusion into the study (a minimum of five employees in each quarter for five years); to calculate company turnover rates and employment growth rates; to calculate wage rates; and to calculate average employee tenure with the business. These in turn allowed us to benchmark companies to their industry by calculating industry claims rates and turnover rates by size of employer; employment growth rates and employee average tenure. Employers on this list were then contacted and requested to complete a survey. In order to include only drivers in the sampling frame for the employee survey, we pulled all employee SSNs for the most recent year from the eligible employers. Under a data-sharing agreement these SSNs were submitted to the state Department of Licensing (DOL), which matched them to their list of current Commercial Drivers License holders. DOL attached address information for the drivers and returned the driver file. Individuals on this list were subsequently contacted to administer the drivers' survey. The administrative data, both from the workers' compensation claims and from the employment security databases, provided valuable context for the survey responses by allowing results to be broken out by employer size, employment growth rate and average driver tenure. Results from the first phase of this study were recently published (Spielholz et. al.; 2008).

Using SESA Data for Measuring the Economic Impact of Injury

Economic losses that result from work-related injury goes far beyond the direct covered medical costs, vocational rehabilitation expenditures, pensions and wage-replacement costs that comprise the direct cost of a workers' compensation claim. Employers bear a portion of the loss in the form of indirect cost including production interruption, accident investigation, and the recruiting and training of replacement workers. There is also the long-term loss of earnings, which extends beyond the period of workers' compensation wage benefits, and is related to the worker's injury-related loss of function or skills. Several studies have used data on earnings from state employment security departments for injured workers and compared them to matched control groups (Boden and Galizzi, 1999; Reville, 1999). One technique for estimating long-term earnings loss is to use multiple regression analysis to compare the earnings of workers with accepted time-loss claims to those of workers with medical-only claims. Covariates such as age, gender, and industry, drawn from the claims record, are included in these models in order to isolate the separate impact of injury type on earnings (Boden and Galizzi, 1999). These studies show that workers who have time-loss injuries are likely to experience substantial income losses that continue long after their wage-replacement benefits end (Reville et al., 2001). More recently this method was extended to estimate the impact of carpal tunnel syndrome on long-term workers' earnings and return to work (Foley et. al, 2007). This study compared quarterly earnings records from the Wage database of the SESA program for workers in Washington State who filed claims in 1993 or 1994 for CTS to those of workers with upper-extremity fracture claims or with medical-only dermatitis claims. Multivariate regression was used to isolate the effect of injury type on earnings from that of other potential predictors. Information from claim records were linked by Social Security Number (SSN) to claimants' quarterly earnings records as reported by their employers to the Washington State Employment Security Department (ESD). In this way, quarterly earnings profiles for all 9,305 claimants were assembled, beginning with the first quarter of 1990 and continuing through the fourth quarter of 2001. Calendar quarters were converted to quarters before/after the quarter of injury to define time elapsed since injury when looking across claimants injured in different calendar quarters. In addition to providing claimants' quarterly earnings, ESD wage records provided information on the claimants' employer at the time of injury. This included the total employment of the establishment as well as the four-digit Standard Industrial Classification (SIC) code. Finally, this database allows us to track breaks in the claimants' employment history over the forty quarters of the study period. The results showed that CTS claimants recover to about half of their pre-injury earnings level relative to that of comparison groups after six years; they also endured periods on time-loss three times longer than claimants with upper extremity fractures, and multiple quarters of reduced hours after initial return to work. Cumulative excess loss of earnings of the 4443 CTS claimants was estimated to be $145 million to $210 million over six years, a loss of $32,000 to $47,000 per claimant.

Further Potential Uses of SESA data

Many opportunities exist to extend the use of linkages between workers' compensation data and SESA data. These include the following:

Compare claims rates for claimants working over 40 hours per week (or 500 hours per quarter) to those working less than forty hours using SESA hours data from the Wage file.

Employees in rural locations (using address information on the Employer file) could face greater difficulty accessing workers compensation healthcare providers. This may result in slower return to work than in more densely provider-served areas.

Compare claims rates for workers in "income/tenure" stable employment to "income/tenure" unstable" employment.

Post-injury, compare return to work and earnings recovery for workers returning to same employer, or in same industry, as opposed to workers returning to work at different employers and industries. Stratify these results by employer size and geography.

Calculate seniority and turnover statistics by industry and employer. Stratify by geographies and by size of employer. Create industry benchmarks at NAICS levels.

References

1: Reville RT, Bhattacharya J, Sager Weinstein LR. New methods and data sources for measuring economic consequences of workplace injuries. Am J Ind Med. 2001 Oct;40(4):452-63. PubMed PMID: 11598994.

2: Boden LI, Ozonoff A. Capture-recapture estimates of nonfatal workplace injuries and illnesses. Ann Epidemiol. 2008 Jun;18(6):500-6. PubMed PMID: 18083542.

3: Bureau of Labor Statistics. Quarterly Census of Employment and Wages Program Overview. Website:http://www.bls.gov/cew/cewover.htm (Viewed August 2009).

4: Spielholz P, Cullen J, Smith C, Howard N, Silverstein B, Bonauto D. Assessment of perceived injury risks and priorities among truck drivers and trucking companies in Washington State. J Safety Res. 2008;39(6):569-76. PubMed PMID: 19064041.

5. Boden LI, Galizzi M. 1999. Economic consequences of workplace injuries and illnesses: lost earnings and benefit adequacy. Am J Ind Med. 36:487-503.

6. Reville RT. 1999. The impact of a disabling workplace injury on earnings and labor force participation. In The Creation and Analysis of Employer-Employee Matched Data. 1999. Elsevier Science.

7. Reville RT, Bhattacharya J, Sager Weinstein LR. 2001. New Methods and Data Sources for Measuring Economic Consequences of Workplace Injuries. Am J Ind Med 40:452-463.

8. Foley M, Silverstein B, Polissar N. The economic burden of carpal tunnel syndrome: long-term earnings of CTS claimants in Washington State. Am J Ind Med. 2007 Mar;50(3):155-72. PubMed PMID: 17216630.

Reconfiguring a Workers' Compensation Database for Epidemiologic Analysis*

Arthur Oleinick, MD, JD, MPH, University of Michigan

Introduction

The Bureau of Labor Statistics annual Survey of Occupational Injuries and Illnesses (BLS SOII) provides invaluable population-based estimates of the incidence of work-related injuries and illnesses by industry (http://www.bls.gov/iif/ohssum.htm). In addition, for the subgroup with days away from work, detailed injury characteristic information is provided by both industry and occupation (http://www.bls.gov/iif/oshcdnew.htm). Unfortunately, the nosology in the Occupational Injury and Illness Classification Manual (OIICM) used to classify the nature of the injury and the part of body affected (http://www.bls.gov/iif/oshoiics.htm) relies on a lay narrative that is then coded by a lay coder. Findings based on such coding can only approximate[1] those that can be obtained by using the standard clinical diagnostic coding reflected in the International Classification of Diseases- Clinical Modification[2] (ICD9-CM).

One result is that it is difficult to identify articles using BLS SOII data in Ovid Medline© so that population-based comparisons of the incidence and outcome of work injuries with identical injuries occurring in other settings is generally precluded. A recent keyword search on keywords reflecting three different federally supported databases provides documentation for this effect. A keyword search of the phrases "SEER (NCI Surveillance, Epidemiology and End Results Centers- http://seer.cancer.gov)," "Framingham heart- NHLBI Framingham Community Study- http://www.framinghamheartstudy.org/index.html) and "Bureau of Labor Statistics" yielded 2792, 1029 and 214 articles, respectively.

The present report suggests that the medical data available in workers' compensation databases can be used to benchmark and enrich the population-based data reported in the BLS SOII.

Methods

The software for this project (Copyright, in part, by the University of Michigan) uses Microsoft's Visual Basic for Applications™ (VBA). From this investigator's perspective, the major strength of this programming language is that the principal investigator can specify the database extract algorithm and then check the output at each step of the resultant PowerPoint™ file by checking the Excel™ file produced by that step, including those files produced by Access's™ ability to combine relational databases. Thus, an investigator who is competent in Microsoft's Office Suite™, but who is not a professional-level programmer, can vouch for the successful transcription of the database extract algorithm.

A data extract was obtained from the Ohio Bureau of Workers' Compensation (OBWC) relational database for workers in the for-hire transportation industry (predominantly employees of trucking firms) who were injured in the study period of 1997-1999 and followed through March of 2002. In Ohio, workers' compensation is provided by a state agency, except those firms with more than 500 employees who qualify for self-insurance.[3] In the 2000 census, Ohio had a civilian employed population of 5.4 million people or about 4.2% of the U.S. total (http://www.census.gov/main/www/cen2000.html) and approximately 70% of the workforce was insured by OBWC.[4] After study approval by the University of Michigan and OBWC, the data extract was provided in nine text files.

Since all work occurs as a result of the combination of movement around major joints and

---

* Supported, in part, by NIOSH Grant R01 OH03804-01A2

involvement of the senses, the basic decision was to create a functional classification system based on ICD9-CM codes. The ICD9-CM taxonomy was reviewed and some 3,400 potential work-related diagnoses were identified, without restriction as to their origin in a particular industry or occupation. Using the diagnostic code and a data element in the OBWC relational database called "Body Object" that provided additional information on anatomic localization, these codes were assigned to the following functional classes: head, back, shoulder, elbow, hand/wrist, carpal tunnel syndrome, hip, knee and ankle/foot. In addition, three general groups were identified: burns, toxic exposures and herniae. No attempt was made to group soft tissue injuries of the thorax or abdomen or psychiatric diagnoses. This software module was completed. Approximately 900 diagnoses were available in the data extract for injured employees of the for-transportation industry (largely truck firm employees) prepared by the OBWC.

A second software module was developed to classify diagnoses within a functional area by severity. Severity classification was based on the nature of the underlying pathology and the usual clinical course for the diagnosis. Permanent tissue changes were rated as more severe than acute tissue changes and diagnoses requiring chronic medical management more severe than diagnoses requiring only a limited number of medical contacts. Severity classifications are complete for the back, shoulder, elbow and knee. Multiple injuries within a functional area were classified by means of a subroutine that produced a bivariate distribution for the two most severe injuries. Together, the completed functional areas account for almost half the diagnoses in the data extract.

A third module incorporated information on injury comorbidities because such comorbidities may identify specific accident patterns or affect outcome. Comorbidities were characterized in terms of their proximity to the index injury. Thus, for back injuries, the proximal comorbidities included injuries of the head, shoulder and hip.

Results

Table 1 gives the functional area for diagnoses in paid claims (about 13% represent a second claim during the 3-year study interval). In almost 81% of injuries all diagnoses were confined to a single functional area, with slightly more than one diagnosis per claim. However, these claims account for only 67% of all diagnoses. Another group of some 2000 claims represent more complicated injuries with injuries to multiple areas yielding an additional 6000 diagnoses. The result is a dramatic increase in the count of injuries in several anatomic areas, in particular the shoulder, elbow, back, hip and knee. Note that 90% of claims can be completely classified by the proposed taxonomy and that these claims account for 85% of the medical information available in the files.

Table 2 shows the two-way distribution of first and second diagnosis for first back injuries during the study period where the most serious back diagnosis is classified as a strain (one of the data extract files provided data on previous claims but we did not get to the software module that would have incorporated this information). Claims in this diagnostic group, for example, had no diagnosis of a displaced intervertebral disc ("herniated disc") but could have an additional diagnosis of contusion or laceration. While 77% of back sprains were confined to a single anatomic area, the remaining 23% have multiple areas involved. In this group of sprains/strains, workers with cervical and thoracic sprains are most likely to have multiple areas of back involvement. Transport accidents account for a disproportionate share of injury events when multiple back sprains are diagnosed, with the exception of cervical sprains where there is little difference in the fraction attributed to transport accidents regardless of whether other back area sprains are diagnosed.

Table 3 shows the injury comorbidities associated with back sprains by whether the back sprains involves one, or more, back areas. Roughly one of eight injured workers with a back sprain confined to a single anatomic area (Part A) also have injury comorbidities except for those with cervical sprains where almost half have such comorbidities. The distribution of comorbidities in those with a cervical sprain also differs with a much higher fraction of Back-Other (including internal injuries of the thorax and abdomen) and Back-Proximal (head, shoulder and hip). In workers with multiple back sprains (Part B) the pattern is similar although the fraction with injury comorbidities among groups other than the cervical sprain group is higher with about one in four injured workers having comorbidities outside the back. In part, the high fraction of comorbidities in workers with cervical sprains again reflects the role of transport accidents in this industry group.

In addition to any studies focused on specific functional areas, Table 4 compares the fraction of the study population of first injuries during the study period with a diagnosis of displaced intervertebral disc ("herniated disc") with the fraction reported by the BLS SOII for various populations.

While the fraction of injuries attributed to dislocations by the BLS SOII data on the left of the table all vary around 0.5%, the fraction based on clinical diagnostic coding in the OBWC data is eight-nine times as large. There are two likely explanations for this difference- first, the lay narrative is inadequate for this level of medical detail and, second, the lay narrative is prepared before the diagnostic evaluation can be completed and so tends to allocate these to the more generic back sprain group.

This lack of specificity for medical detail in the BLS SOII is unfortunate because serious medical conditions are likely to account for a disproportionate share of health care and its costs so that population-based studies of this condition by industry and/or occupation might identify interventions that would reduce both the incidence and outcome for these serious conditions.

Discussion

It is clear that workers' compensation data can be used to supplement, benchmark and enrich the population-based data available in the BLS SOII. The availability of complete medical diagnostic information for work injuries in some state workers' compensation databases makes possible a three-dimensional view of work injuries- functional area affected, severity and injury comorbidities- that more clearly identify the effect of the injury event. Such a multi-faceted view is unavailable in the BLS SOII which mandates a preference for coding each injury to the single most severe injury if it can be determined.[5] In addition, using the medical diagnostic information, studies of medical care use, rehabilitation and outcome can be compared to controlled studies of care and outcome for the same condition in the literature as a way to identify how job characteristics might facilitate or hinder rehabilitation or return-to-work.

Table 1. Functional Area for Paid Claims, Ohio BWC For-hire Transportation Firms, 1997-1999

Functional Area	Number of Paid Claims	
	All diagnosis(es) in same area	Any diagnosis in area
Shoulder	1287	2361
Elbow	718	1291
CTS	44	109
HW	4615	5230
Back	4890	6596
Head	2060	2725
Hip	467	986
Knee	1651	2444
AF	2426	2783
Hernia	174	216
TDB	369	405
TE	306	339
Total	19007	21212^1 $(22216)^2$
% paid claims	80.90%	90.3% (94.6%)
# diagnoses	22753	29007^1 $(>31280)^2$
% all diagnoses	66.60%	84.9% (>91.6%)

1. The number represents the number of claims in which all diagnoses are classified to the functional areas listed.
2. The number in parentheses represents the number of claims in which at least one of the diagnoses is classified to the functional areas listed.

Table 3. The Fraction of First Back Injuries Having Injury Comorbidities Involving Other Functional Areas During the Study Period in the For-hire Transportation Industry in Ohio, 1997-1999.

	Proximity of Other Functional Areas to the Back Injury Area Diagnosis					
	Distal	Proximal	Proximal & Distal	Back Only	Other	Total
Single back area						
Lumbosacral	4.06%	5.98%	1.58%	87.71%	0.68%	887
Sacral	4.96%	5.67%	2.13%	86.52%	0.71%	141
Cervical	7.30%	19.65%	4.53%	55.42%	13.10%	794
Thoracic	2.63%	10.50%	1.67%	84.73%	0.48%	419
Lumbar	4.65%	4.59%	1.08%	89.08%	0.61%	1483
Coccyx				100.00%		1
Grand Total	4.86%	8.83%	2.07%	80.97%	3.28%	3725
≥2 back areas						
Lumbosacral (100%)[1]	6.83%	9.35%	7.19%	73.74%	2.88%	278
Sacral (87.4%)	5.56%	18.52%	1.85%	72.22%	1.85%	54
Cervical (90.4%)	11.09%	14.60%	4.62%	63.03%	6.65%	541
Thoracic (61.8%)	8.54%	5.49%	1.22%	82.93%	1.83%	164
Lumbar (77.9%)	10.53%	12.28%	1.75%	73.68%	1.75%	57
Grand Total	9.32%	12.05%	4.47%	69.68%	4.47%	1095

[1] The proportions in parentheses are the fraction of cases in a particular anatomic area captured by the first back sprain diagnosis categorized and reflects the software design.

Table 2. First and Second ICD9-CM Codes for Work-related First Back Sprains During Study Period by Anatomic Area among For-Hire Transportation Employees, Ohio, 1997-1999

Second back area diagnosis[2]	Back injury area of first diagnosis indicating a sprain/strain[1]						
	Unspecified	Lumbosacral	Sacral	Cervical	Thoracic	Lumbar	Coccyx	Total
None		887	141	794	419	1483	1	3725
Sacral		28						28
Cervical	1	125	15					141
Thoracic		46	7	308				361
Lumbar		46	28	212	152			438
Coccyx			1			2		3
Lesser back injuries	0	33	3	21	12	55	0	124
Total	1	1165	195	1335	583	1540	1	4820
Any in area		1165	223 (195+28)	1476	944	1978	4	
% multiple areas involved[3]		23.86%	36.77%	46.21%	55.61%	25.03%		

[1] Excludes approximately 700 back sprain cases, most of which (442) had the back sprain as a second injury in the period and some (102) had one or more non-localizing back sprain diagnoses.
[2] The absence of a row for lumbosacral sprains represents an artifact created by the way the software collected the diagnoses for categorization.
[3] 4.4% of claims had three, or more, back diagnoses listed of any kind.

Table 4. BLS SOII Estimates of Proportions of Dislocations of Back and Neck[1] among All Occupations and among Truck Drivers with First Injury Diagnosed as Displaced Intervertebral Discs ("Herniated Discs") and among Truck Drivers in the Ohio For-hire Industry (1997-1999).

BLS[2]	
All occupations, U.S. (1996 – 2000)	0.47-0.35%
All occupations, Great Lakes Region (2000)	0.52%
Truck Drivers, U.S. (1996 – 2000)	0.66-0.34%
Truck Drivers, Great Lakes Region (2000)	117 [3]
OBWC (1997 – 1999) – Truck Drivers	
ICD9-CM Acute Disc Codes	ICD9-CM 722.0, 722.2, 722.7* (722.8*[4])
Proportion of all injuries	3.85-4.50%[5]
OBWC (1997 – 1999) – All occupations	
Wage compensation (>7 disability days)	89.5%[6]

1. Category includes slipped, ruptured or herniated disc
2. Data for the Great Lakes states (Minnesota, Wisconsin, Illinois, Indiana and Michigan) were provided from a special run of BLS SOII data. Ohio does not participate in the BLS SOII system.
3. Number of cases. Three of five Great Lakes' states listed no dislocations.
4. One case had this code as sole diagnosis.
5. If the 27% of cases where occupation was not identified broke the same as those with occupation (unlikely since we used four data sets to identify truck drivers), the range would be 2.81-3.27%. The range reflects differing degrees of certainty in ascertaining the occupation of truck drivers.
6. Percentage of all 458 first injuries first displaced intervertebral disc injury cases in all occupations, Ohio BWC.

References

1. Lipscomb HJ, Cameron W, Silverstein B. 2008. Back injuries among union carpenters in Washington State, 1989-2003. Am. J. Ind. Med. 51:463-474.

2. National Center for Health Statistics (NCHS). International Classification of Diseases, Ninth Revision, Clinical Modification (ICD9-CM). Washington, DC; Government Printing Office; 2002 (HE 22.41/2:2002)

3. Ohio Revised Code (Anderson). 2007. §4123.35(B)(1)(a)

4. Ohio Bureau of Workers' Compensation (OBWC).BWC Library. 2005. Facts and figures: 2002-2004 Fact Sheet URL: http://www.ohiobwc.com/home/current/FactFigLnks.asp

5. http://www.bls.gov/iif/oshoiics.htm, Section 2.1, Rules 1.3 and 1.4.

The Use of Workers' Compensation Data to Identify and Track Workplace Risk and the Effectiveness of Preventative Measures

Edward Bernacki, MD, MPH, Johns Hopkins University and Hospital

Workers' compensation data can be useful to track cause-specific injury incidence and to evaluate the effectiveness of prevention measures (Bernacki and Guidera, 1998, Bernacki et al., 1999, Bernacki and Tsai, 2003). The JHU workers' compensation program is self-insured and is administered by the university Health, Safety and Environment Department (HSE) in compliance with State of Maryland regulations. The program covers all employees of Johns Hopkins University, Johns Hopkins Hospital and a small number of affiliates. Program components are listed in Table 1.

Table 1. Johns Hopkins Workers' Compensation Program Components

- Safety and Industrial Hygiene
- Healthcare Provider Network
- In-House Case Management
- In-House Claims Payment and Management
- Integration of Components Utilizing Web Based Software

The Integrated Workers' Compensation Claims Management System (IWCCMS) at Johns Hopkins University (JHU) history and general structure were described by Bernacki & Tsai (2003). The primary function of IWCCMS is to make payments to claimants and physicians. The system and support program have evolved substantially since 1992 when some of the first changes were instituted. Some of the key changes are listed in Table 2.

Table 2. Implementation Steps for Integrated Workers' Compensation Claims Management System at Johns Hopkins University

- Environmental monitoring/surveillance program established.
- Small network (6+) of clinically skilled OEM and surgical specialists with knowledge of the workers' compensation system developed.
- Nurse case manager (NCM) hired to facilitate the diagnostic and treatment process.
- Information transferred among physicians, safety professionals, nurse case manager and supervisors via meetings and a claims management software system (ICMS).
- Case Manager insures claim status known at all times.
- Transitional Duty Pool created.
- Case Manager and Software System supports (rather than hinders) physician's ability to diagnose, treat and return individuals to productive work.
- All participants facilitate the continuous assessment and correction of work areas where accidents occur (accident investigation).
- Preoccupation with timely payments to claimants and physicians.

Portions of the integrated system information may be accessed by department staff to track performance measures and to evaluate the effectiveness of hazard interventions. For example, the number of ergonomics surveys was followed over a five-year period when the rate of upper extremity work-related musculoskeletal disorders (UEWMSD) decreased (Figure 1) and the need for surgical interventions all but disappeared. The steps in the ergonomic survey are listed in Table 3.

Table 3. Elements of Ergonomic Survey

- Document repeated, sustained and forceful exertions.
- Document awkward postures.
- Summarize ergonomic stressors and risk factors.
- Provide information on ergonomic principles on a one-to-one basis.
- Document and report results of ergonomic survey.
- Monitor the implementation of corrective actions.

The annual number of UEWMSD reached a minimum in 1999 (Figure 2). The number has generally been increasing over the past decade while the rate showed little change from 2001 - 2009. The increased numbers of disorders are due to greater numbers of employees covered by IWCCMS. The number of surgical treatments for the disorders has been near zero since the mid-1990s.

Costs for the Johns Hopkins workers' compensation program have increased since the start of the tracking system in 1992 (Figure 3). Yet the losses per $100 payroll have been reduced over this period and that ratio has remained relatively consistent since the late 1990's.

References

Bernacki EJ, Guidera JA. (1998) The effect of managed care on surgical rates among individuals filing for workers' compensation. J Occup Environ Med ;40(7):623-31.

Bernacki EJ, Guidera JA, Schaefer JA, Lavin RA, Tsai SP (1999) An Ergonomics Program Designed to Reduce the Incidence of Upper Extremity Work Related Musculoskeletal Disorders. J Occup Environ Med.. Volume 41(12) 1032-1041

Bernacki EJ, Tsai SP (2003) Ten years' experience using an integrated workers' compensation management system to control workers' compensation costs. J Occup Environ Med. 45(5):508-16.

Figure 1. Interventions to Prevent Work-related Musculoskeletal Disorders

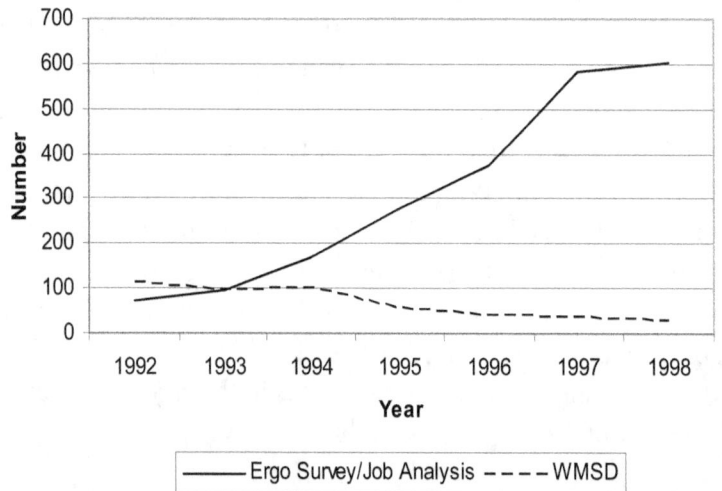

Figure 2. Upper Extremity Work-related Musculoskeletal Disorders over 18 Year Period

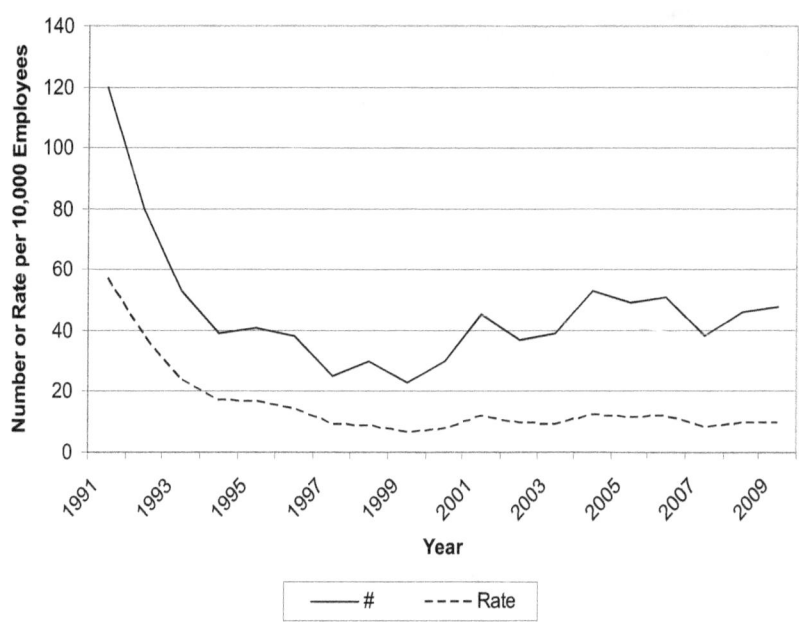

Figure 3. Johns Hopkins Workers' Compensation Costs Compared with Payroll

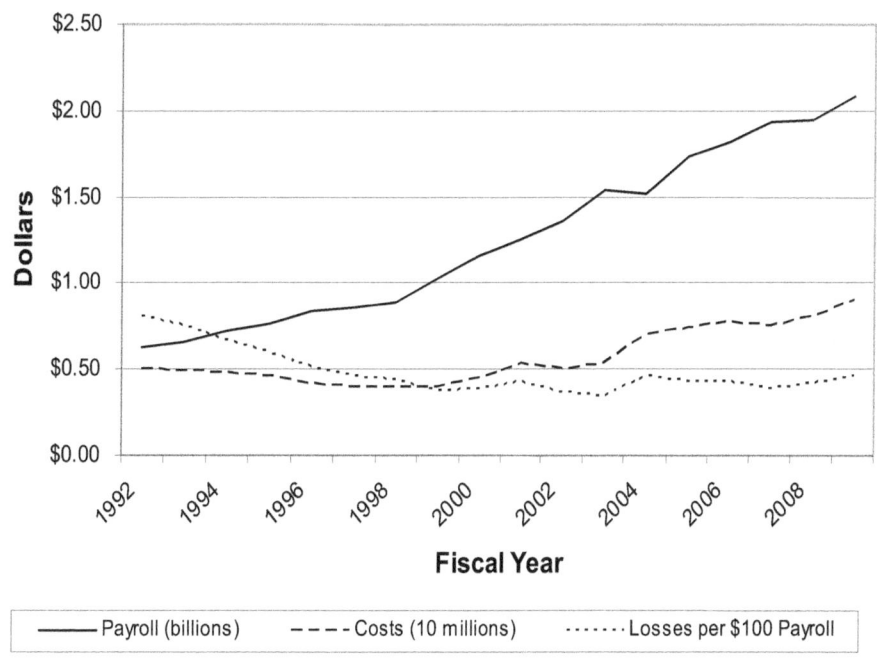

Data Linkage for Prevention: Traumatic Injuries in Construction

Linda Forst, MD, MPH, Lee Friedman, PhD, University of Illinois at Chicago School of Public Health

Introduction

Epidemiological surveillance is the systematic collection, analysis and dissemination of health data for planning, implementing, and evaluating public health programs. Occupational surveillance focuses on monitoring the health of working populations and exposure hazards in the workplace (1). The four essential components of an occupational surveillance system include: 1) gathering information on adverse health events and exposure circumstances; 2) distilling and analyzing the data; 3) disseminating information to interested parties; and 4) intervening on the basis of the evidence provided by the data to alter the factors that produced the hazards and adverse health outcomes. How can occupational surveillance be used? It enables public health officials, businesses, researchers, enforcers, and other stakeholders to: 1) become familiar with the magnitude and distribution of occupational illnesses and injuries; 2) monitor trends over time; 3) identify emerging injury and exposure problems; 4) flag specific cases or situations for follow up investigations; 5) set intervention priorities; and 6) evaluate intervention activities.

Data linkage is a methodology that entails merging individual cases across two or more databases. "Deterministic" methods utilize exact matches on designated data elements, usually relying on identifying data (e.g., name, social security number); "probabilistic" matching is necessary when identifiers are not available. With data linkage, it is possible to create a new dataset that includes more variables than those that are available in either database, alone. This methodology also allows the researcher to assess the quality of data entry by comparing the information that has been entered for data elements that are common to both databases and to fill in data that is missing in one of the original datasets if it has been entered in the other. Finally, the merging of several databases can increase overall case capture for specific injury or disease sentinels.

For this workshop on use of workers' compensation data, the overall goal of this project was to explore the potential of linking cases in workers' compensation claims and the state trauma registry in Illinois. We selected "construction" injuries to test this methodology. Specific objectives were to:

1) gain experience with the data linkage process;
2) calibrate our expectations of the number of matches from two separate databases; and
3) describe acute, severe construction injuries in Illinois.

Methods

There are approximately 60,000 workers' compensation claims filed in Illinois each year. Workers, and sometimes their attorneys or other worker advocates, file claims when there is a dispute, or the expectation of a dispute, regarding compensation after a work-related injury. We obtained a dataset of all workers' compensation claims in Illinois for the year 2005 from the Illinois Workers' Compensation Commission. There was no information available regarding "economic sector" (SIC code). We, therefore, obtained a list of all construction companies in Illinois from Manufacturers News, Inc (manufacturersnews.com). We matched the company names in the WC dataset with the names in the Manufacturer's News list. We also made a list of key words related to construction, and conducted a key word search of companies listed in the WC dataset to find additional construction cases. We merged the two case selection methods to craft a single dataset of construction injury claims in 2005. To identify duplicates, we matched all of these cases on four variables (name, sex, date of birth, date of accident), and purged 1.4% of the cases due to duplication. A more detailed description of our methodology is reported in another manuscript (Friedman and Forst, in press).

The Illinois Trauma Registry captures every injured individual that gets hospitalized in a level one or two designated trauma centers in Illinois. In 2005, there were 47,091 cases reported. After de-duplication (matching on name, gender, date of birth, date of injury, and race/ethnicity), we ended with 45,978 cases of both occupational and non-occupational cases.

We conducted a deterministic linkage, matching on name, sex, and date of birth. We found 264 cases that linked between the two databases. We determined the number of cases we would expect, and we did a descriptive analysis of the newly derived dataset.

Results
We had a sample of 2736 construction injury cases from the WC Claims in 2005, representing approximately 4% of the claims that year. These cases linked with 264 cases in the Illinois Trauma Registry. We utilized the Bureau of Labor Statistics (BLS) Survey of Occupational Injuries and Illnesses (SOII) in 2005 to determine the number of construction injury cases we might expect. In 2005, 9.5% of injuries in Illinois were in the construction sector. There are approximately 2900 occupational cases per year in the Illinois Trauma Registry. If we multiply 2900 X 0.095, we would expect 276 construction cases in the linked database. Our observed/expected was 264/276, or 95.7%. Table 1 shows demographics and median compensation by gender, age group, marital status, and body site of injury. Table 2 shows median compensation against measures of severity on admission, discharge outcome, and permanent disability. Median compensation-- including medical care, lost time, and permanent partial or total disability payments--were greatest in married people, back injuries, when more than one body site was affected, in 25-44 year olds, and in death cases. Upper extremity injuries was the most injured body site, though back injuries were almost twice as costly in terms of compensation payments per injury. Falls were the most common mechanism of injury, followed by motor vehicle crashes and assaults (Table 3).

Discussion and Next Steps
This is the first published report of linkage between a workers' compensation claims database and state trauma registry data. Unlike the BLS SOII which relies on employer reporting, the claims database depends on worker reporting (by filing a claim), and the trauma registry entails automatic reporting by health care systems of every case that is cared for in a trauma center. Automatic reporting from health care systems allows for capture of cases that may not be reported by employers or injured workers. Because these cases landed in trauma centers, these are likely to be the most severe injuries (in Illinois, deaths outside of the hospital or in the emergency department do not get reported to the trauma registry); illnesses, such as poisonings or acute respiratory events, would not be cared for in a trauma unit, and are, thus, systematically missed in Illinois trauma registry data. Under-capture of occupational illness and injury cases is expected from every database, and there is no way to know how many cases were missed in this one. Our analysis indicates that the linkage procedure identified a high proportion of the expected cases.

Cases that are triaged to trauma units are the most severe injuries, some of which could (and did) result in death. The proportion of specific causes of injury reflects the reported causes of death in construction according to the BLS SOII: in 2005, "falls" were the mechanism of injury in 32.0% of fatalities, "transportation" was 28.4%, and "contact with objects" was 20.0% (2). These proportions are similar to the overall causes of hospitalization in trauma units for occupational injuries, as well—"falls" (34.8%), "machinery" (18.1%) and "motor vehicle crashes" (11.2%) (3). The results of the analyses, after data linkage procedures, are what would be expected: ages 25-44 have the highest number and proportion of injuries and also resulted in the greatest median compensation, reflecting that this group has the highest employment in construction and that they are likely to do the most hazardous tasks; there are few women in this group; upper extremity injuries are the most common, but injuries of the back and spine are the most costly; the median compensation increases with percent temporary and percent permanent disability. Median compensation for place of discharge and injury severity was not logical—discharge to home being more costly than discharge to rehab or a nursing home-- though the small number of cases and the inability to garner more details about the cases make these findings difficult to interpret.

This project demonstrates the potential for conducting complex analyses that can inform intervention priorities and areas to examine further. Our next steps will be to conduct a more rigorous data linkage project using newly available software for more years of data. With a larger dataset, we will be able to conduct more informative data analyses, controlling for confounders. For example, we could correlate exposure risk with long- and short-term outcomes, severity of injury with prognosis, and mechanism, severity, and diagnosis with cost.

Our next steps are to work with the Illinois Workers' Compensation Commission to improve reporting so that the problem of missing data is minimized and the quality of the data is better. We also hope to work with the Illinois Workers' Compensation Commission to communicate directly with the Occupational Safety and Health Administration (OSHA) and with other stakeholders. There is potential to make data linkage ongoing and in real-time, to report specific sentinels directly from the database to OSHA, and to report to stakeholders who can utilize summary results for their missions. Further, we can use data linkage to prioritize statewide interventions--developing policy initiatives and educational programs—and to evaluate preventive activities. Finally, we can share this methodology among the states to improve occupational surveillance and prevention, nationwide.

References

1. Halperin WE, Frazier TM. Surveillance for the Effects of Workplace Exposure. Annual Rev Public Health 1985;6:419-32.

2. CPWR. The Construction Chartbook. Leading Causes of Fatal and Non-fatal Injuries in Construction. http://cpwr.com/pdfs/CB%20 4th%20Edition/31_50%20Safety%20and%20 Health.pdf#page=11; accessed 11/14/09

3. Friedman L, Forst L. Occupational Injury Surveillance of Traumatic Injuries in Illinois using the Illinois Trauma Registry, 1995-2003. J Occup Environ Med 2007;49(4):401-10.

Table 1. Median compensation by demographics in traumatically injured construction workers in Illinois, 2005

	N (%)	Median Compensation
Gender		
Male	258 (97.7%)	$11,144
Female	6 (2.3%)	$12,908
Marital Status		
Single	104 (39%)	$10,000
Married	154 (58.3%)	$13,649
Age		
16-24	23 (8.7%)	$4500
25-34	78 (29.5%)	$11,281
35-44	75 (28.4%)	$18,829
45-54	66 (25.0%)	$4980
55-64	21 (8.0%)	$2221
>65	1 (0.4%)	$2170
Body Site		
Upper Extremities	83 (31.4%)	$12,141
Lower Extremities	55 (20.8%)	$13,058
Back and Spine	34 (12.9%)	$22,833
Head and neck	14 (5.3%)	$2,985
Torso	5 (1.9%)	$4,200
Internal	1 (0.4%)	---
Mult extr. Unspec	87 (33.0%)	$6,180

Table 2. Number of cases and median compensation in traumatic construction injuries, by measures of severity, discharge outcome, percent permanent disability and percent partial disability, Illinois 2005.

Characteristic	N (%)	Median Compensation
Measures of severity		
Mean days in hospital	4.84±8.24	~
Sent to intensive care unit	62 (23.5%)	$13,649
Mean days in ICU	6.15±8.81	~
Put on ventilator	28 (10.6%)	$11,762
Mean days on ventilator	7.57±8.65	$14,795
In hospital fatality	13	$56,163
In hospital fatality rate	4.92%	~
Mean injury severity score	8.38±7.79	~
ISS=16-24	35 (13.3%)	$14,795
ISS≥25	13 (4.9%)	$5,332
Discharge Outcome		
Death	30 (11.4%)	$56,163
Discharged home	203 (76.9%)	$12,500
Rehab or acute care facility	10 (11.4%)	$10,233
Nursing home/res. Facility	11 (4.2%)	$4,051
Other	7 (2.7%)	$21,598
Percent Perm Disability		
No disability	151 (57.2%)	$0
1-25%	72 (27.3%)	$17,343
26-50%	32 (12.1%)	$33,063
51-100%	9 (3.4%)	$40,000

Table 3. Causes of severe traumatic injury in construction in Illinois, 2005

Fall	94 (35.6%)
Motor Vehicle Crash	79 (29.9%)
Assault	37 (14.0%)
Struck by/caught between	17 (6.4%)
Cutting or piercing instrument	11 (4.2%)
Machinery	8 (3.0%)
Fire or caustic substance	6 (2.3%)
Electrocution	2 (0.8%)
Animal bite	2 (0.8%)
Suffocation	1 (0.4%)
Other	7 (2.7%)

Workers' Compensation Coverage by State

Ishita Sengupta and Virginia P. Reno, National Academy of Social Insurance (NASI)*

Consistent with the goal of this workshop, we aim to draw insight from our experience at the National Academy of Social Insurance (NASI) with workers' compensation data. We understand that the goal is to consider how workers' compensation data might be used for occupational health and safety surveillance.

The purpose of NASI's annual report on workers' compensation benefits, costs, and coverage is to provide a benchmark to facilitate policymaking and comparisons with other social insurance and employee benefit programs. Workers' compensation pays for medical care, rehabilitation, and cash benefits for workers who are injured on the job or who contract work-related illnesses. It also pays benefits to families of workers who die of work-related causes. Each state has its own workers' compensation program.

Need for NASI's Report

The lack of uniform reporting of states' experiences with workers' compensation makes it necessary to piece together data from various sources to develop estimates of benefits paid, costs to employers, and the number of workers covered by workers' compensation. The lack of a federally mandated data system means that states vary greatly in the data they have available to assess their programs. Some states have excellent systems. Others can provide little or no information.

The audience for the Academy's reports on workers' compensation includes journalists, business and labor leaders, insurers, employee benefit specialists, federal and state policymakers, and researchers in universities, government, and private consulting firms[1]. The Academy's data are not used for surveillance.

The data report is produced with the expert advice of the Academy's Study Panel on National Data on Workers' Compensation, chaired by John F. Burton, Jr. Members of the expert panel are listed at the back of this paper. The Academy staff and its expert advisors are continually seeking ways to improve the report and to adapt estimation methods to track new developments in the insurance industry and in workers' compensation programs.

NASI Estimates

NASI estimates workers' compensation medical benefits and cash wage-replacement benefits for each state and for the nation as a whole. We also estimate employer costs, the number of workers' covered by law, and the aggregate wages of those covered workers. We draw on other sources to report trends in workers' compensation claims and frequency of workplace injuries.

In 2007, workers' compensation covered an estimated 131.7 million workers, an increase of 1.1 percent from the 130.3 million workers covered in 2006. Total

---

*Ishita Sengupta is the Workers' Compensation Research Associate and Virginia P. Reno is the Vice-President for Income Security Policy at the National Academy of Social Insurance in Washington D.C. Its mission is to promote understanding of how social insurance contributes to economic security and a vibrant economy.

[1] The U.S. Census Bureau publishes the data in the Statistical Abstract of the United States; the National Safety Council uses the data in Injury Facts; and the Employee Benefit Research Institute uses these data in its reference work, Fundamentals of Employee Benefit Programs. The Social Security Administration publishes the data in its Annual Statistical Supplement to the Social Security Bulletin. The Centers for Medicare & Medicaid Services use the data to estimate and project health care spending in the United States. The National Institute for Occupational Safety and Health uses the data to track the cost of workplace injuries. In addition, the International Association of Industrial Accident Boards and Commissions (the organization of state and provincial agencies that administer workers' compensation in the United States and Canada) uses the information to track and compare the performance of workers' compensation programs in the United States with similar systems in Canada.

wages of covered workers were $5.9 trillion in 2007, an increase of 5.6 percent from 2006 (Sengupta et al 2009). Workers' compensation benefits paid in 2007 were $55.4 billion, which included $27.2 billion for medical care and $28.3 billion for cash compensation to workers. Total benefits rose by 2.0 percent over 2006. Employer costs for workers' compensation in 2007 were $85.0 billion, a decline of 2.7 percent from the prior year.

Coverage Rules

Every state except Texas requires almost all private employers to provide workers' compensation coverage (U.S. DOL 2006). In every state an employee not covered by workers' compensation insurance or an approved self-insurance plan is allowed to file suit claiming the employer is liable for his or her work-related injury or illness.

Some states exempt from mandatory coverage certain categories of workers, such as those in very small firms, certain agricultural workers, household workers, employees of charitable or religious organizations, employees of some units of state and local government, and railroad employees who are covered by other arrangements. Employers with fewer than five workers are exempt from mandatory workers' compensation coverage in Alabama, Arkansas, Florida, Georgia, Michigan, Mississippi, Missouri, New Mexico, North Carolina, South Carolina, Tennessee, Virginia, and Wisconsin. The rules for agricultural workers vary among states. In eleven states (in addition to Texas), farm employers are exempt from mandatory coverage altogether. In other states, coverage is compulsory for some or all farm employers. The largest groups of workers who are not covered under either unemployment insurance or workers' compensation are self-employed individuals who have not incorporated their businesses.

Method for Estimating Coverage

NASI seeks to count the number of workers who are legally required to be covered under state laws. We do not attempt to estimate compliance with state laws. To the extent that employers fail to comply with state coverage laws or misclassify workers as independent contractors, our estimates will overstate coverage. To the extent that the relatively few employers for whom coverage is voluntary do provide coverage of their workers, our estimates would understate coverage.

Because no national system exists for counting workers covered by workers' compensation, the number of covered workers and their covered wages must be estimated. We start with the number of workers in each state who are covered by unemployment insurance (UI) (U.S. DOL 2008). Almost all of U.S. wage and salary workers are covered by UI (NASI 2002). U.S. employers who are required to pay unemployment taxes must report quarterly to their state employment security agencies information about their employees and payroll covered by unemployment insurance. These data are a census of U.S. workers who are covered by unemployment insurance. We subtract from UI coverage estimates of the number of workers and the amount of wages that are not required to be covered by workers' compensation because of exemptions for small firms, farm employers, and because coverage for employers in Texas is voluntary.

Using these methods we estimate that in 2007, 97.3 percent of all UI–covered workers and wages were covered by workers' compensation. Table 1 in the appendix shows NASI estimates of covered workers' in each state.

Use of Coverage Data for Surveillance

To count covered workers for surveillance purposes, presumably one would want to estimate compliance with state workers' compensation laws. In-depth studies at the state level would be needed to document compliance and we know of no national system that exists for this purpose. A 2007 study in New York found that actual coverage was less than legally required coverage because of failure to report employees or misclassifying them as independent contractors (Fiscal Policy Institute 2008, Greenhouse 2008). An in-depth review of states' approaches to regulating coverage of independent contractors was conducted by a joint working group of the National Association of Insurance Commissioners and the International Association of Industrial Accident Boards and Commissions--the association of state workers' compensation agencies (NAIC/IAIABC 2008). The group's review of research found that misclassification of employees as independent contractors is a large and growing problem for workers' compensation and other programs.

Claims or Workplace Injuries

NASI does not try to make independent estimates of workers' compensation claims. As a proxy, we use reports by the Bureau of Labor Statistics (BLS) from its Survey of Occupational Injuries and Illnesses for private industry. A total of 1.2 million non-fatal workplace injuries or illnesses that required recuperation away from work beyond the day of the incident were reported in 2007. The reported incidence of such injuries or illnesses has declined steadily since 1992 – from 3.0 per 100 full-time workers in 1992 to 1.4 in 2004 and to 1.2 in 2007. We also use the Census of Fatal Occupational Injuries that is compiled by the BLS, which shows a gradual decline in fatal occupational injuries since 1992[2].

To corroborate the BLS trends in private-industry workplace injuries we examine trends in workers' compensation claims compiled by the National Council on Compensation Insurance (NCCI). These data show a similar downward trend in workers' compensation claims. The number of workers' compensation claims for temporary-total disability per 100,000 insured workers declined by nearly half between 1992 and 2004 (Sengupta et al 2009).

The NCCI data are for the subset of workers whose employers purchase private insurance and who live in the 36 states for which NCCI provides statistical and ratemaking advisory services. Missing from these data are workers covered by employers who insure themselves for workers' compensation. The self-insureds account for about one-fourth of benefit spending in 2007. Also missing from the NCCI data are workers who live in the so-called "non-NCCI" states, such as California, Michigan, New York, Ohio, Pennsylvania and Washington[3].

Obtaining data that are consistent across self-insured employers, privately insured employers, and those who buy insurance from state funds is a major challenge. We know of two efforts that are underway that aim to improve consistency of data across states. First, the IAIABC has an initiative underway to promote more consistent reporting of workers' compensation claims electronically. Its efforts to promote more consistent reporting rest on voluntary participation by the states with funding provided by individual state legislatures as they see fit. A second activity underway is housed at the Workers' Compensation Research Institute (WCRI 2009). It has initiated surveys in individual states to track the experience of injured workers and to provide more consistent measures of the performance of workers' compensation programs across states.

Concluding Observations

With the help of an outstanding group of workers' compensation experts, we at NASI have been able to put together consistent and comprehensive data on workers' compensation benefits, costs, and coverage. Surveillance of occupational health and safety is not the purpose of our data. If surveillance is the mission, it would seem necessary to find a way to align regulatory oversight and data reporting requirements with that mission.

---

[2] The number of work-related death declined from 6,217 in 1992 to 5,657 in 2007. In 2001, an additional 2,886 deaths at work were attributed to the September 11 attacks.

[3] Other non-NCCI states are Delaware, Massachusetts, Minnesota, New Jersey, North Dakota, Texas, West Virginia, Wisconsin and Wyoming.

References

Fiscal Policy Institute. 2007. "New York State Workers' Compensation: How Big is the Coverage Shortfall?" A Fiscal Policy Institute Report, www.fiscalpolicy.org

Greenhouse, Steven. 2008. "Dozens of Companies Underpay or Misreport Workers, State Says." The New York Times, February 12.

NAIC-IAIABC. 2008. " An Overview of Workers' Compensation Independent Contractor Regulatory Approaches", NAIC/IAIABC Joint Working Group of the Workers' Compensation (C) Task Force.

National Academy of Social Insurance (NASI). 2002. "Workers' Compensation Coverage: Technical Note on Estimates." Workers' Compensation Data Fact Sheet, no. 2. Washington, DC: National Academy of Social Insurance.

Sengupta I, Virginia Reno and John F. Burton Jr. 2009. Workers' Compensation: Benefits, Coverage, and Costs, 2007. Washington, DC: National Academy of Social Insurance.

TDI (Texas Department of Insurance). 2008. Biennial Report of the Texas Department of Insurance to the 81st Legislature, Division of Workers' Compensation. http://www.tdi.state.tx.us/wc/ regulation/roc/pdf/wc0904est.pdf

U.S. Department of Labor (U.S. DOL). 2006. Employment Standards Administration. Office of Workers' Compensation Programs. 2006 State Workers' Compensation Laws. Washington, DC: U.S. DOL.

U.S. Department of Labor (U.S. DOL). 2008. Employment Standards Administration. Office of Workers' Compensation Programs. ES 202 Report. Washington, DC: U.S. DOL.

U.S. Small Business Administration (U.S. SBA). Office of Advocacy. 2008. Employer Firms, Establishments, Employment, and Annual Payroll by Firm Size and State, 2006. Washington, DC: U.S. SBA.

WCRI (Workers' Compensation Research Institute). 2009. CompScope™ Benchmarks, 9th Edition, January, Cambridge, MA.

Table 1. Documenting Workers' Compensation Coverage Estimates, 2007 Annual Averages

States	UI Covered Jobs[a]		Workers' Compensation Exemptions			WC Covered Jobs	WC as a % of UI
	Total	Private, NF[b]	Small Firm[c]	Agriculture	Texas		
Alabama	1,900,205	1,586,252	71,955	5,728	-	1,822,522	95.9
Alaska	293,990	233,882			-	293,990	100.0
Arizona	2,595,406	2,234,894			-	2,595,406	100.0
Arkansas	1,152,994	968,901	26,856	7,264	-	1,118,874	97.0
California	15,394,639	13,014,420			-	15,394,639	100.0
Colorado	2,240,513	1,922,611			-	2,240,513	100.0
Connecticut	1,666,470	1,434,570			-	1,666,470	100.0
Delaware	417,944	363,483			-	417,944	100.0
D.C.	487,301	448,890			-	487,301	100.0
Florida	7,818,063	6,803,332	313,953		-	7,504,110	96.0
Georgia	3,981,342	3,391,525	90,426		-	3,890,916	97.7
Hawaii	594,108	500,874			-	594,108	100.0
Idaho	647,823	532,679			-	647,823	100.0
Illinois	5,782,011	5,043,457			-	5,782,011	100.0
Indiana	2,868,833	2,492,221		10,810	-	2,858,023	99.6
Iowa	1,467,394	1,240,875			-	1,467,394	100.0
Kansas	1,332,447	1,103,410		8,381	-	1,324,066	99.4
Kentucky	1,764,091	1,497,495		3,950	-	1,760,141	99.8
Louisiana	1,837,327	1,523,063			-	1,837,327	100.0
Maine	588,210	500,302			-	588,210	100.0
Maryland	2,422,400	2,082,136			-	2,422,400	100.0
Massachusetts	3,185,031	2,820,720			-	3,185,031	100.0
Michigan	4,125,157	3,542,854	94,287		-	4,030,870	97.7
Minnesota	2,654,671	2,301,405			-	2,654,671	100.0
Mississippi	1,109,288	890,536	44,342	7,890	-	1,057,056	95.3
Missouri	2,664,949	2,286,474	110,282		-	2,554,667	95.9
Montana	423,281	353,809			-	423,281	100.0
Nebraska	900,631	752,917			-	900,631	100.0
Nevada	1,267,618	1,129,428		2,181	-	1,265,437	99.8
New Hampshire	622,435	542,711			-	622,435	100.0
New Jersey	3,899,781	3,341,428			-	3,899,781	100.0
New Mexico	791,206	628,836	19,801	8,419	-	762,986	96.4
New York	8,427,180	7,103,486			-	8,427,180	100.0
North Carolina	3,999,628	3,366,215	90,890		-	3,908,738	97.7
North Dakota	331,988	275,242		2,416	-	329,572	99.3
Ohio	5,229,964	4,542,944			-	5,229,964	100.0
Oklahoma	1,489,454	1,211,137			-	1,489,454	100.0
Oregon	1,698,837	1,426,009			-	1,698,837	100.0
Pennsylvania	5,548,936	4,917,786			-	5,548,936	100.0
Rhode Island	470,274	416,914			-	470,274	100.0
South Carolina	1,861,707	1,557,105	60,159	6,504	-	1,795,044	96.4
South Dakota	380,982	319,694			-	380,982	100.0
Tennessee	2,696,396	2,330,378	93,120	5,276	-	2,598,000	96.4
Texas	10,047,097	8,486,904			2,411,303	7,635,794	76.0
Utah	1,183,684	1,020,381			-	1,183,684	100.0
Vermont	297,491	249,584			-	297,491	100.0
Virginia	3,516,428	2,994,322	79,699		-	3,436,729	97.7
Washington	2,857,266	2,355,510			-	2,857,266	100.0
West Virginia	684,192	568,999			-	684,192	100.0
Wisconsin	2,751,715	2,384,708	58,070		-	2,693,645	97.9
Wyoming	270,467	214,781			-	270,467	100.0
U.S. non-Federal	132,641,24	113,252,489	1,153,839	68,819	2,411,303	129,007,283	97.3
	2,726,304	-				2,726,304	100.0
U.S. Total	135,367,54	113,252,489	1,153,839	68,819	2,411,303	131,733,587	97.3

[a] UI-covered employment reported in the ETA-202 data produced by the United States Bureau of Labor Statistics (U.S. DOL, 2008).
[b] Data not available for 2007, used the 2006 data.
Source: National Academy of Social Insurance estimates.

National Academy of Social Insurance
Workers' Compensation Data Study Panel

John F. Burton, Jr., Chair
Professor Emeritus
Labor Studies & Employment Relations, School of Management & Labor Relations, Rutgers University

Marjorie Baldwin
Professor, W. P. Carey School of Business
School of Health Management and Policy
Arizona State University

Peter S. Barth
Professor of Economics, Emeritus
University of Connecticut

Christine Baker
Executive Officer
CHSWC

Keith Bateman
Vice President, Workers' Compensation
Property Casualty Insurers Association of America

Leslie Boden
Professor, School of Public Health
Boston University

Aaron Catlin
Deputy Director, National Health Statistics Group,
Office of the Actuary, Centers for Medicare and Medicaid Services

James N. Ellenberger
Former Deputy Commissioner
Virginia Employment Commission

Shelby Hallmark
Director, Office of Workers' Compensation Programs
U.S. Department of Labor

Jay S. Himmelstein, M.D.
Professor, Family Medicine and Community Health
Chief Health Policy Strategist,
Center for Health Policy and Research,
University of Massachusetts Medical School

Douglas J. Holmes
President, UWC -Strategic Services on Unemployment
and Workers' Compensation

H. Allan Hunt
Senior Economist
W.E. Upjohn Institute

Kate Kimpan
Vice President, Workers' Compensation Programs
Dade Moeller & Associates

Gregory Krohm
Executive Director
International Association of Industrial Accident Boards and Commissions

Barry Llewellyn
Senior Divisional Executive, Regulatory Services
National Council on Compensation Insurance, Inc.

Eric Nordman
Director of Research
National Association of Insurance Commissioners

Mike Manley
Research Coordinator
Oregon Department of Consumer and Business Services

Frank Neuhauser
Research Faculty
University of California, Berkeley

Robert Reville
Director, Institute for Civil Justice
RAND

John Ruser
Assistant Commissioner for
Safety, Health and Working Conditions
U.S. Bureau of Labor Statistics

Emily A. Spieler
Dean and Edwin W. Hadley Professor of Law
Northeastern University School of Law

Robert Steggert
Vice President, Casualty Claims
Marriott International, Inc.

Alex Swedlow
Executive Vice President/Research
California Workers' Compensation Institute

Richard A. Victor
Executive Director
Workers' Compensation Research Institute

Alex Wasarhelyi
Project Officer, Social Security Administration

Benjamin Washington
Statistician, National Health Statistics Group,
Office of the Actuary, Centers for Medicare and Medicaid Services

William J. Wiatrowski
Associate Commissioner
Office of Compensation & Working Conditions
U.S. Department of Labor, Bureau of Labor Statistics

Summary of Workshop Discussion: Occupational Health and Safety Surveillance Using Workers' Compensation Data

Barbara Silverstein, Washington State Department of Labor and Industries
Emily Spieler, Northeastern University
David Utterback, NIOSH
Teresa Schnorr, NIOSH
Tom Leamon, Harvard University
Letitia Davis, Massachusetts Department of Public Health

The Workshop was convened to discuss opportunities for collaboration in the analysis of workers' compensation (WC) data in order to help reduce the risks of occupational injuries and illnesses. The participants were a diverse group of people – those employed in WC systems, representatives of industry, unions and insurance/reinsurance companies, public health practitioners, regulators, government representatives from several federal and state public health agencies, and academic researchers.

These different groups see the utility of WC data from very different vantage points. While insurers and WC agencies may use WC claims experience primarily for ratemaking purposes, public health researchers mine the data to understand population trends in work-related injuries and illnesses, including incidence, lost work time, costs, and shifts in attribution. Although there is great interest in using WC data for these public health purposes, these data are not generally available to individual researchers or organizations: many large data collections have restrictive confidentiality agreements of some kind; some data sets developed by rating agencies can be prohibitively expensive.

One of the key themes to emerge during the conference was "seek first to understand before being understood." Participants considered a number of sources of data potentially available for surveillance purposes including: state and federal WC agencies; state insurance agencies; National Academy of Social Insurance (NASI) reports; employers; insurance carriers (including state funds); reinsurance companies; rating agencies (e.g. National Council on Compensation Insurance (NCCI)); and the Bureau of Labor Statistics (BLS) and recognized the various sensitivities of each source and the need to accommodate them in any work using these data.

Below is a brief summary of extensive dialogue that took place throughout the workshop as the participants discussed the challenges of using WC data for public health purposes. This summary organizes the comments and ideas of the participants by general strengths, barriers, and concerns of insurers, followed by opportunities for linkage of data, some suggestions for further exploration and some next steps.

Strengths of WC Data

WC data that are systematically collected for employer, insurer and medical uses have properties that are consistent with occupational health and safety surveillance needs:

1. WC is nearly universal. All states except Texas require employers' to carry WC coverage. Exclusive state fund data may be most comprehensive.

2. Information that is not available from other sources on incidence and rates of injuries may be collected in WC systems, including data stratified by state, industry, establishment, and corporation.

3. Ongoing collection of WC claims data is usually closer to real-time than BLS surveys. Yet, there is nevertheless a lag time even for WC first reports to appear in data systems, and the lag varies by jurisdiction.

4. High value information is often tracked in WC systems on lost work-time, temporary/permanent, partial/full disabilities along with reported causation and the portions of total costs that are disbursed to claimants and health care providers.

5. Medical treatments, their costs and outcomes are associated in WC data with specific injuries and classes of injuries.

6. Disability cause, duration, and cost data are generally quite complete.

7. Narrative text in WC claim files can be used to identify hazards and illustrate their essential factors.

Limitations for Use of WC Data for Public Health

These major strengths of WC data are matched by equally significant and very real barriers:

1. There are significant and not well understood variations among the state WC systems so that it is extremely difficult to combine and compare data. Regulations for which employers are mandated to provide coverage, the occupational illnesses and the severity of injuries that are reportable, and many other factors vary among states.

2. Numerous conditional filters affect record contents. These also vary by jurisdiction. Variations exist in the definition of accepted claims, who files claims, medical treatment guidelines, willingness of physicians to treat a claimant, administrative process and litigation rules, among others.

3. Proprietary ownership, state laws and state practices, and federal and state privacy and confidentiality rules can inhibit data sharing and analysis.

4. Many costs of occupational injuries and illnesses are not fully covered by the WC system and are therefore not included in the WC data. For example, under state laws wage replacement is often partial, and under some circumstances work-related disabilities may become the responsibility of other forms of social insurance.

5. Employers can manage WC claims and costs through approaches that affect both initial reporting of injuries and illnesses as well as duration of disability.

6. The actual numbers of injuries and illnesses that may be classified as work-related for public health purposes may be greater than those recorded as paid claims due to the legal definition of compensable injuries and illnesses and administrative rules for claims within WC systems.

7. Occupational diseases can be absent from WC data due to a variety of reasons such as failure of the employee to file the claim due to lack of knowledge of work-relatedness, lack of sufficient evidence that the disease is work-related and procedural legal rules (e.g. time limitations for filing claims).

8. Many factors influence decisions by injured workers to file a WC claim, resulting in incomplete data. Among others, these include: language barriers; perceived or real concerns

regarding adverse effects on the relationship with the employer; lack of knowledge about WC rights; peer pressure; and immigration status. Research suggests, however, that underreporting is more prevalent for less severe injuries[1].

9. Data in individual records often are incomplete and otherwise vary in quality within and among potential sources.

10. Discrepancies between first reports of injuries and later data for the same claim are common and have varying lags.

11. Fraud by all parties in the system – health care providers, employers, employees, carriers – detrimentally affects the quality of data.

12. Various stakeholders have different vantage points that may bias interpretations. For example, decisions on what to report may be interpreted quite differently, depending on vantage point: some might see the absence of an injury report as an inappropriate exclusion of a claim, while others might see it as the proper exclusion of a non-compensable injury.

WC Carriers Concerns with Providing Data for Public Use

WC systems are designed to provide payments in compensable claims to claimants and health care providers, and not to develop public health data. Carriers and others have concerns that create barriers to use of these data for alternative purposes:

1. Data systems and their contents are proprietary.

2. WC insurance carriers' competition for customers may be impaired through partial or complete disclosure of company specific data.

3. There is a concern that the risks of legal actions may increase with public access to the data.

4. Interpretation of data element definitions and the nuanced meaning of industry terms could result in misinterpretation and inappropriate use of data.

5. Variability in how data are collected and stored may lead to misguided interpretations when data from multiple sources are combined.

6. Cost of providing data is not trivial. Requestors are often unaware of the time required to create, run and check queries of data collected and organized for other purposes.

7. Ownership and control of the data may become problematic after they are released for public health surveillance.

8. Complexities and costs of changing coding systems are large. For example, moving from American National Standards Institute (ANSI) code structures to the Occupational Injury and Illness Classification System (OIICS) is enormously expensive and time-consuming.

9. Functionality and quality of the data and its organization may be limited due to the systems employed for collection and retrieval.

10. Coding inconsistencies and gaps may be due to reasons that are not always a matter of record. For example, WC data that are used for managing claims payments may be more complete and accurate than other data elements such as cause and nature of injury, which are important for public health surveillance.

11. Business interests in data are not always consistent with research or public health interests. For example, carriers are primarily concerned with managing claims costs as well as safety and loss control while the public health community focus is primary and secondary prevention.

Potential Uses of WC Data

Despite the potential difficulties with WC data access and representativeness, many participants thought further collaborative efforts could be productive in many ways for all stakeholders.

1. Stakeholders in business, labor, and government have an interest in the WC data regarding disability and medical treatment costs and trends.

2. Analysis of incidence, cost and lost time data can be used to make a business case for prevention and to provide employers with additional information and incentives to reduce claims through interventions.

3. Cost-benefit evaluations would be more informative for interventions through the examination of the financial, medical, and disability data from WC systems.

4. Carriers and companies may use consolidated WC data for benchmarking.

5. Data on frequency of events within industries or by region can initiate prevention efforts and help ensure more effective resource allocation.

6. Systematic review of claims can help identify new hazards within industries or individual employers. Different rates or severities among employee groups may trigger identification of new hazards or conditions.

7. Faster analysis of frequency or incidence of claims is possible with WC data since they are often closer to real time than other data sources such as those from BLS.

8. WC data can identify elements of cost for occupational injuries even though it is much less useful for occupational diseases, particularly for those with long latency periods.

9. Medical treatment and cost information for specific injuries and conditions can lead to better comparisons across larger populations of workers.

10. It may be possible to study the impacts of differences in financial incentives for workplace safety by evaluating the relationship between the type of insurance, e.g., retrospective rating program, and its effects on claims experience.

11. Descriptive information on individual workplaces and individual workers may be used to identify intervention opportunities.

Data Linkages

One of the important outcomes of this workshop was the identification of potential benefits of combining WC data with other sources to inform decision makers about the magnitude and cost of occupational illnesses and injuries.

1. More complete estimates for the actual count of work-related injuries and illnesses may be developed through linking WC, BLS and state registries.

2. The total real cost of occupational injuries and illnesses could be estimated more accurately by linking data from WC, BLS, unemployment

insurance, and Social Security Disability Insurance.

3. Newer technologies such as electronic medical records could lead to better tracking of the full consequences of occupational injuries and illnesses when combined with WC and other existing medical data.

4. Medical treatment effectiveness can be evaluated through combinations of data from WC, disability systems and medical data systems.

5. Effectiveness of regulatory requirements can be evaluated through review of WC data with cooperation of the Occupational Safety and Health Administration (OSHA), Mine Safety and Health Administration (MSHA) and state OSHA programs.

6. Review of mandated employer reports and disability data that are collected by WC, private and public disability insurance and the Social Security Administration could lead to better public health intervention strategies.

Ideas and Approaches

Participants generally agreed that WC data could be useful for occupational health and safety surveillance despite the barriers and difficulties. A variety of approaches should be tested before national implementation plans are contemplated. The ideas discussed during the workshop include:

1. Potential parties must identify effective approaches that encourage WC carriers, academia and government to collaborate.

2. The focus should initially be pilot efforts in a few states to determine what data are most useful for tracking purposes at the employer, carrier and public health levels and as a prelude to coordinating and standardizing methods.

3. Capacity in state health and labor departments should be increased to support work with stakeholders that use WC data systems for evaluating emerging exposures and workplace injuries and illnesses and to provide feedback to employers, carriers and public health departments.

4. The full burden of occupational illness and injury can be captured by looking at the long-term consequences of work related injuries, e.g., the odds of osteoarthritis due to a work-related fracture of the hip may be greater than the odds of osteoarthritis for those without a history of work-related hip fracture.

5. Homogeneous groups of industries and regions should be identified for comparisons among injury and illness data sets.

6. Benchmarking could be provided by trade associations or others that use combined data sets to allow employers and carriers to compare their programs to others in the same industry.

7. If a national disability database were established, WC data could be used to examine the contributions from occupational disorders.

8. A joint venture is possible for BLS and the International Association of Industrial Accident Boards and Commissions (IAIABC) to work on standardizing reporting.

9. A WC check box should be included and completed on medical and hospital forms for systems such as Medical Expenditures Panel Survey (MEPS) and Healthcare Cost and Utilization Project (HCUP) to better capture work-related conditions. BLS should capture in SOII whether

a particular injury or illness case involved a hospitalization or emergency department visit. These data elements would allow BLS estimates to be better compared to WC and other data.

10. Some suggested that consideration should be given to resurrecting the Supplemental Data Systems* conducted in the past by the BLS.

Next Steps

The workshop discussion summary and the articles in this proceedings form a basis for further work on the topic of WC data for surveillance purposes. Participants identified two important next steps in order to develop the potential synergies identified above.

1. Representatives from private sector insurance carriers, trade associations, professional organizations, NCCI, reinsurers, labor unions, public sector regulators, including WC administrations and state insurance commissioners, and federal research agencies, including BLS and NIOSH, should develop strategies to address the following through additional meetings or other forms of collaboration:

   a. Develop approaches to improve the availability and value of U.S. WC data for public health purposes in order to improve health and reduce costs;

   b. Identify specific gaps in knowledge about the use of WC data to address public health issues that might be addressed through new research collaborations;

   c. Develop strategies to increase the transparency and availability of all data consistent with privacy, confidentiality and proprietary issues; and

   d. Develop organizational capacity and secure funds to support research and analysis across jurisdictions.

2. Federal and state regulatory, public health and research agencies (BLS, NIOSH, OSHA, MSHA, State Health and Labor Departments) and WC agencies and organizations should work to assess the potential uses of WC data and to integrate individual electronic medical records and WC data systems.

In conclusion, workshop participants suggested that WC data could be used to great advantage while protecting the interests of individuals, employers, and carriers - both private and public (state and federal) plans. The extensive discussions identified a variety of collaborations that could improve our ability to utilize WC data for surveillance purposes in order to identify trends, sentinel events and potentially higher risk work arrangements and to test strategies that are intended to reduce work-related illnesses and injuries.

Reference
1. Rosenman KD; Gardiner JC; Wang J; Biddle J; Hogan A; Reilly MJ; Roberts K; Welch E (2000). J Occup Environ Med; 42(1):25-34.

The Supplemental Data System (SDS) was carried out by the Bureau of Labor Statistics and participating States in the late 1970s and 1980s. Based on surveys of workers identified from WC reports, the SDS collected detailed information on specific types of workplace injuries and illnesses, such as those related to servicing equipment, falls from elevations, and back injuries related to lifting. Some of the data elements in the SDS are now obtained on a broader and more statistically sound basis by the Survey of Occupational Injuries and Illnesses in the collection of information on case circumstances and worker characteristics.

Workshop Participants

Amick III, Benjamin C., Ph.D., Professor
University of Texas School of Public Health
Institute for Work & Health

Azaroff, Lenore, Sc.D.
Research Professor
University of Massachusetts Lowell

Baker, Christine, M.A.
Executive Officer
Commission on Health and Safety and Workers' Compensation

Bang, Ki Moon, Ph.D., M.P.H.
Senior Research Epidemiologist
CDC/NIOSH

Bateman, Keith, J.D.
Vice President, Workers Compensation
Property Casualty Insurers Association of America

Bernacki, Edward, M.D., M.P.H.
Executive Director, Health, Safety and Environment
Johns Hopkins University

Bhattacharya, Anasua, Ph.D.
Economist Fellow
CDC/NIOSH

Boden, Leslie, Ph.D.
School of Public Health
Boston University

Bodine, Edward F., Ph.D.
Analyst
U.S. Government Accountability Office

Boiano, James, M.S., CIH
Senior Industrial Hygienist
CDC/NIOSH

Bonauto, David, M.D., M.P.H.
Associate Medical Director, SHARP Program
Washington State Department of Labor and Industries

Bracy, Michelle
Senior Analyst
Government Accountability Office

Brewer, Shelley, Dr.P.H., CSP
Chemical Loss Control Specialist
ChemPlan, Inc.

Burton, John F., Ph.D.
Professor Emeritus
Rutgers University

Bushnell, Tim, Ph.D., M.P.A.
Economist
CDC/NIOSH

Butler, Richard J., Ph.D.
Professor
Brigham Young University

Carson, Sandra G., RN, COHNS
Vice President, Crisis Management
Sysco Corporation

Castillo, Dawn, M.P.H.
Chief, Surveillance and Field Investigations Branch
CDC/NIOSH

Conway, George, M.D., M.P.H.
Director, Agriculture Forestry and Fishing Program
CDC/NIOSH

Courtney, Theodore, M.S., CSP
Director, Center for Injury Epidemiology
Liberty Mutual Research Institute for Safety

Davis, Letitia, Sc.D, Ed.M.
Director, Occupational Health Surveillance
Massachusetts Department of Public Health

Dembe, Allard E, Sc.D.
Chair and Associate Professor
The Ohio State University, College of Public Health

Duncan, John C., M.A.
Director
Department of Industrial Relations, California

Foley, Mike, Ph.D.
Senior Economist, SHARP Program
Washington State Department of Labor and Industries

Forst, Linda, M.D., M.P.H
Professor
University of Illinois – Chicago, School of Public Health

Frumin, Eric, M.S.
Health and Safety Director
Workers United/SEIU/Change to Win

Galassi, Thomas
Directorate, Technical Support and Emergency Management
Occupational Safety and Health Administration

Gallahan, John
Statistician
Bureau of Labor Statistics

Getahun, Abay Asfaw
Senior Economics Fellow
CDC/NIOSH

Gillen, Matt, M.S., CIH
Construction Program Coordinator
CDC/NIOSH

Gittleman, Janie, Ph.D.
Associate Director, Safety and Health Research
CPWR – The Center for Construction Research and Training

Goddard, Keith
Director, Directorate of Evaluation and Analysis
Occupational Safety and Health Administration

Hall, Keith, Ph.D.
Commissioner
Bureau of Labor Statistics

Harrison, Robert, M.D., M.P.H
Chief, Occupational Health Surveillance and Evaluation Program
California Department of Public Health

Headley, Tanya, M.S.
Health Communications Specialist
NIOSH/CDC

Howard, John, M.D., M.P.H.
Director
CDC/NIOSH

Hudock, Stephen, Ph.D., CSP
Team Leader, Human Factors and Ergonomics Research
CDC/NIOSH

Jackson, Larry L., Ph.D.
Supervisory Epidemiologist
CDC/NIOSH

Jinnett, Kimberly, Ph.D.
Research Director
Integrated Benefits Institute

Kole, Jennifer
Occupational Safety and Health Specialist
Occupational Safety and Health
Administration

Leamon, Tom B., Ph.D.
Adjunct Professor of Occupational Safety
Harvard School of Public Health

Leigh, J. Paul, Ph.D.
Department of Public Health Services
University of California, Davis

Lin, Mei-Li, Ph.D.
Executive Director, Research & Statistical Services
National Safety Council

Lotz, W. Gregory, Ph.D.
Director, Division of Applied Research & Technology
CDC/NIOSH

Lyons, Chris
Senior Analyst
Government Accountability Office

Marucci-Wellman, Helen, Sc.D., M.S.
Research Scientist
Liberty Mutual Research Institute for Safety

Mendeloff, John, Ph.D.
Director, RAND Center for Health and Safety in the Workplace
Professor, University of Pittsburgh

Michaels, David Ph.D, MPH
Assistant Secretary of Labor (Designee)
for the Occupational Safety and Health Administration

Neuhauser, Frank
Researcher, Survey Research Center
University of California, Berkeley

Newman, Katharine, Ph.D.
Economist
Bureau of Labor Statistics

Noy, Ian, Ph.D. CPE
Director
Liberty Mutual Research Institute for Safety

Oleinick, Arthur, M.D., J.D., M.P.H.
Associate Professor, School of Public Health
University of Michigan

Pana-Cryan, Rene, Ph.D.
Economics Program Coordinator
CDC/NIOSH

Pierce, Brooks, Ph.D.
Research Economist
Bureau of Labor Statistics

Reno, Virginia, Ph.D.
Vice President for Income Security
National Academy of Social Insurance

Rosenman, Kenneth D., M.D.
Professor of Medicine
Michigan State University

Ruser, John W., Ph.D.
Assistant Commissioner
Bureau of Labor Statistics

Schmid, Frank, Ph.D.
Director and Senior Economist
National Council on Compensation Insurance

Schmidt, Dave
Director, Office of Statistical Analysis
Occupational Safety and Health Administration

Schneider, Scott, M.S.
Director, Occupational Safety and Health
Laborers' Health and Safety Fund of North America

Schnorr, Terri, Ph.D.
Director, Division of Surveillance, Hazard Evaluations and Field Studies
CDC/NIOSH

Schultheiss, Victor
Head of Center of Competence for Workers' Compensation Insurance
Munich Reinsurance Company

Seabury, Seth, Ph.D.
Economist
RAND Corporation

Seidner, Adam L., M.D., M.P.H
Medical Director
Travelers Workers Compensation Claim

Seminario, Peg, M.S.
Safety and Health Director
AFL-CIO

Sengupta, Ishita, Ph.D.
Workers' Compensation Research Association
National Academy of Social Insurance

Sestito, John, J.D., M.S.
Surveillance Coordinator
CDC/NIOSH

Shaffer, Tom
Chief, Division of Safety and Health Statistics
Bureau of Labor Statistics

Shuford, Harry, Ph.D.
Chief Economist
National Council on Compensation Insurance

Silverstein, Barbara, M.S.N, Ph.D., M.P.H
Research Director, SHARP Program
Washington State Department of Labor and Industries

Sokas, Rosemary, M.D., M.O.H.
Director, Office of Occupational Medicine
Occupational Safety and Health Administration

Souza, Kerry, Ph.D.
Epidemiologist
CDC/NIOSH

Spieler, Emily, J.D.
Dean and Professor of Law
Northeastern University School of Law

Steggert, Robert
Vice President, Casualty Claims
Marriott International, Inc.

Stewart, Mary L., ARM-P, CPCU
Director, R&D
Public Entity Risk Institute

Stout, Nancy, Ed.D.
Director, Division of Safety Research
CDC/NIOSH

Sygnatur, Eric, Ph.D.
Economist
Bureau of Labor Statistics

Upegui-García, Héctor, M.D.
Senior Consultant
Center of Competence for Workers'
Compensation Insurance
Munich Reinsurance Company

Utterback, David F., Ph.D.
Services Sector Coordinator
CDC/NIOSH

Wiatrowski, William
Associate Commissioner
Bureau of Labor Statistics

Wurzelbacher, Steven, Ph.D., CPE
Research Industrial Hygienist
CDC/NIOSH

Zaidman, Brian, Ph.D.
Senior Research Analyst
Minnesota Department of Labor and Industry

Acronyms and Abbreviations

A&TS	Actuarial & Technical Solutions, Inc
AAFA	Asthma and Allergy Foundation of America
AMA	American Medical Association
AOEC	Association of Occupational and Environmental Clinics
ATS	American Thoracic Society
AWW	Average Weekly Wage
BACs	benzalkonium chlorides
B2B	Business to Business
BLS	Bureau of Labor Statistics
BRFSS	Behavioral Risk Factor Surveillance System
Cal/OSHA	California Occupational Safety and Health Administration
CB	Claims Both
CDC	Centers for Disease Control and Prevention
CDI	California Department of Insurance
CDPH	California Department of Public Health
CDR	Center for Disability Research
CDWC	California Division of Workers' Compensation
CHAMPUS	Civilian Health and Medical Program of the Uniformed Services
CHIS	California Health Interview Survey
CIE	Center for Injury Epidemiology
CM	Claims Medical
COPD	Chronic Obstructive Pulmonary Disease
CPI	Consumer Price Index
CSLB	Contractors State License Board
CSTE	Council of State and Territorial Epidemiologists
CTS	Case Tracking System
DAFW	days away from work
DALYs	disability adjusted life years
DCI	Detailed Claims Information
DFR	Doctors First Report
DIR	Department of Industrial Relations
DLI	Department of Labor and Industries
DLSE	Division of Labor Standards Enforcement
DOL	Department of Labor
DOL	Department of Licensing (WA)
DOSH	Division of Occupational Safety and Health
DWC	Division of Workers' Compensation
ED	emergency departments
EDD	Employment Development Department
EEEC	Economic and Employment Enforcement Coalition

ELP	English language preference
ESD	Employment Security Department
FDLT	first day of lost time
FEIN	Federal Employer Identification Number
FROI	First Report of Occupational Injury
FTE	full-time equivalent
HCUP	Healthcare Cost and Utilization Project
HMO	Health Maintenance Organization
HSA	Health and Safety Associations
HSE	Health, Safety and Environment Department
IAIABC	International Association of Industrial Accident Boards and Commissions
ICD9	International Classification of Diseases
ICD9-CM	International Classification of Diseases-Clinical Modification
IBI	Integrated Benefits Institute
ILO	International Labor Organization
IWCCMS	Integrated Workers' Compensation Claims Management System
IR	OSHA Recordable Incident Rate
IWH	Institute for Work and Health
L&I	Washington State Department of Labor and Industries
LB	Liability Data Mart
LBD	lower back disorders
LEP	limited English proficiency
LEWMSD	lower extremity work-related musculoskeletal disorder
LM	Liberty Mutual
LMRIS	Liberty Mutual Research Institute for Safety
LMWSI	Liberty Mutual Workplace Safety Index
LTD	Long Term Disability
MCDB	Managed Care Database
MEPS	Medical Expenditures Panel Survey
MOL	Ministry of Labor
MSD	musculoskeletal disorders
MSHA	Mine Safety and Health Administration
MSPI	Medical Services Price Index
NAICS	North American Industry Classification System
NASI	National Academy of Social Insurance
NHIS	National Health Interview Study
NHLBI	National Heart, Lung and Blood Institute
NIOSH	National Institute for Occupational Safety and Health
NCCI	National Council on Compensation Insurance
NCEH	National Center for Environmental Health
NCHS	National Center for Health Statistics
NOS	not otherwise specified
NSC	National Safety Council

OA	occupational accident
OBWC	Ohio Bureau of Workers' Compensation
OD	occupational disease
OEM	Occupational and Environmental Medicine
OHS	occupational health and safety
OHSCO	Occupational Health and Safety Council of Ontario
OIICS	Occupational Injury and Illness Classification System
OSH	occupational safety and health
OSHA	Occupational Safety and Health Administration
OT	overtime (pay code)
P2P	person to person
PAR	Participatory Action Research
PCAs	Patient Care Assistants
PCTs	Patient Care Technicians
PERI	Public Entity Risk Institute
PI	prevention index
PPD	permanent partial disability
PTD	permanent total disability
PTO	paid time off (pay code)
QCEW	Quarterly Covered Employment and Wages
Reg	regular (pay code)
RMIS	Risk Management Information System
RTW	return to work
SAIF	State Accident Insurance Fund
SB	Senate bill
SCF	State Compensation Fund
SDS	Supplementary Data System
SEER	Surveillance, Epidemiology and End Result
SENSOR	Sentinel Event Notification System for Occupational Risk
SESA	State Employment Security Agencies
SF	State Fund
SHARP	Safety & Health Assessment and Research for Prevention
SHE(O)	Sentinel Health Event (Occupational)
SIC	Standard Industrial Classification
SLP	Spanish language preferring
SOII	Survey of Occupational Injuries and Illnesses
SPH	Safe Patient Handling
SSN	Social Security Number
STD	short-term disability
STI	Straight time (pay code)
TIRES	trucking injury reduction emphasis on safety
TPA	third-party administrator
TPD	temporary partial disability

TTD	temporary total disability
UBI	Universal Business Identifier
UCFE	Unemployment Compensation for Federal Employees
UEBTF	Uninsured Employers Benefits Trust Fund
UEF	Uninsured Employers Fund
UEWMSD	upper extremity work-related musculoskeletal disorder
UI	unemployment insurance
UNC	University of North Carolina
VBA	Visual Basic for Applications
WC	workers' compensation
WCI	workers' compensation insurance
WCIRB	Workers' Compensation Insurance Rating Bureau
WCMSA	Workers' Compensation Medicare Set-Aside Arrangement
WCIS	Workers' Compensation Information System
WCPR	Workers' Compensation Policy Review
WCRI	Workers' Compensation Research Institute
WHO	World Health Organization
WIC	Washington State Industrial Classification
WRA	work-related asthma
WSIB	Workplace Safety Insurance Board

Using Workers' Compensation Data for Occupational Injury & Illness Prevention
September 22-23, 2009
Washington, D.C.

Tuesday, September 22

8:00   Check In

8:30   Welcome and Introductions of Workshop Organizers
Keith Hall, BLS Commissioner, John Howard, NIOSH Director, Terri Schnorr, NIOSH

8:45   Workshop Overview – Why Use Workers' Compensation Data for Public Health and Protection?
Barbara Silverstein and Tom Leamon

9:15   Panel on History of Workers' Compensation
Les Boden, Boston University, Moderator

A Brief History of Economists' Research on the Effect of Workers' Compensation on Safety and Health
–John Burton, Rutgers University

Historical Perspectives on the Relationship of Workers' Compensation Research to Public Health
–Allard Dembe, Ohio State University

Discussion

10:00 Break

10:15 Panel on Methods
Seth Seabury, RAND, Moderator

Reconciling Workplace Injury and Illness Data Sources
–John Ruser, BLS

Methodological Challenges in the Liberty Mutual Workplace Safety Index: Working Towards a Future Model
–Helen Marucci-Wellman, Liberty Mutual

Harmonizing Existing Databases Counting Workplace Injuries and Illnesses
–Arthur Oleinick, University of Michigan

Workers' Compensation Coverage by State
–Ishita Sengupta, NASI

Discussion

11:15   Panel on Insurance Companies Workers' Compensation Data Use
Tom Leamon, Harvard University, Moderator

Overview of a Workers' Compensation Carrier's Service Sector Data
–Adam Seidner, Travelers

Worker's Compensation Data Utilization at the Liberty Mutual Research Institute for Safety
–Ted Courtney, Liberty Mutual

Pitfalls of Using Employer Records –Does Cleaning the Data Change Effects?
–Shelley Brewer, Chemplan

Discussion

12:15   Lunch (on your own)

1:15   Panel on State and Provincial Workers' Compensation
Peg Seminario, AFL-CIO, Moderator

Using Key Databases in California for Research, Policy and Oversight
– John Duncan, DIR and Christine Baker, CCHSWC

State-based Surveillance and Workers Compensation Data
–Bob Harrison, UC-San Francisco

Ontario use of Workers' Compensation Data for Risk Reduction
–Ben Amick, Institute for Work and Health

Discussion

2:15   Break

2:30   Panel on Public Employee/Self Insured Workers' Compensation Data
        Nancy Stout, NIOSH, Moderator

Benchmarking Public Sector Claims
–Mary Stewart, PERI

Aligned Incentives: Data & Dollars
–Bob Steggert, Marriott

Using Workers' Compensation Data: The Move from Lagging to Leading Indicators
–Sandra Carson, Sysco

Discussion

3:30   Group Discussion
Tish Davis, Massachusetts Department of Public Health, Moderator

5:00   Adjourn

Wednesday, September 23

8:30   Introductions and Announcements

8:45   Past, Present and Future Uses of Some Workers' Compensation Data
Paul Leigh, UC-Davis

9:15   Aggregating Costs and Evaluating Trends
Rene Pana-Cryan, NIOSH, Moderator

Differences among State Laws and Regulations
–Keith Bateman, PCIAA

Estimates of the National Averages of Employee Benefits and Employer Costs for Workers' Compensation
–John Burton, Rutgers University

Learning from Workers Compensation Claims Triangles
–Frank Schmid, NCCI

Discussion

10:15 Break

10:30 Aggregating Costs and Evaluating Trends Continued
Steve Wurzelbacher, NIOSH, Moderator

Assessing Trends in Frequency and Severity and What Drives Those Trends
–Harry Shuford, NCCI

Identifying Vulnerable Populations in Workers' Compensation Data: Temporary Workers and Spanish Language Preference Workers
–David Bonauto, Washington State Department of Labor and Industries SHARP

Discussion

11:15 Opportunities for Prevention
Ben Amick, Institute for Work and Health, Moderator

How to Make Interventions Work: An Insurance Perspective
–Hector Upegui-Garcia, Munich Reinsurance Company

Workers' Obstacles to Filing Claims
–Lenore Azaroff, University of Massachusetts at Lowell

Discussion

12:00 Lunch (on your own)

1:00 Analysis of Linked Data Sets
Janie Gittleman, The Center for Construction Research and Training, Moderator

Lost Work Days under Workers' Compensation and Short-Term Disability Systems
–Kim Jinnett, IBI

Linking of Workers' Compensation and Employment Data
–Mike Foley, Washington State Department of Labor and Industries SHARP

Reconfiguring a Workers' Compensation Database for Epidemiologic Analysis
–Art Oleinick, University of Michigan

The Use of Workers' Compensation Data to Identify Risk in the Workplace and the Effectiveness of Preventative Measures
–Ed Bernacki, Johns Hopkins University

Illinois Workers Compensation and Other Health Data
–Linda Forst, University of Illinois Chicago

Discussion

2:30 Break

2:45 Brainstorming – New and Better Ways to Use Workers' Compensation Data
Barbara Silverstein Washington State Department of Labor and Industries SHARP, and Emily Spieler, Northeastern University, Moderators

Strengths and Limitations in the use of Workers' Compensation Data For Risk Reduction/Prevention

Barriers to Access and Use

Emerging Standards and Reporting Formats

Opportunities or Threats – Potential Impact Electronic Health Records

3:45 Adjourn

www.ingramcontent.com/pod-product-compliance
Lightning Source LLC
Chambersburg PA
CBHW080243180526

45167CB00006B/2393